T0134915

Advanced Sciences and Technologies for Security Applications

Indexed by SCOPUS

The series Advanced Sciences and Technologies for Security Applications comprises interdisciplinary research covering the theory, foundations and domain-specific topics pertaining to security. Publications within the series are peer-reviewed monographs and edited works in the areas of:

- biological and chemical threat recognition and detection (e.g., biosensors, aerosols, forensics)
- crisis and disaster management
- terrorism
- cyber security and secure information systems (e.g., encryption, optical and photonic systems)
- traditional and non-traditional security
- energy, food and resource security
- economic security and securitization (including associated infrastructures)
- transnational crime
- human security and health security
- social, political and psychological aspects of security
- recognition and identification (e.g., optical imaging, biometrics, authentication and verification)
- smart surveillance systems
- applications of theoretical frameworks and methodologies (e.g., grounded theory, complexity, network sciences, modelling and simulation)

Together, the high-quality contributions to this series provide a cross-disciplinary overview of forefront research endeavours aiming to make the world a safer place.

The editors encourage prospective authors to correspond with them in advance of submitting a manuscript. Submission of manuscripts should be made to the Editor-in-Chief or one of the Editors.

More information about this series at http://www.springer.com/series/5540

P. Andrew Karam

Radiological and Nuclear Terrorism

Their Science, Effects, Prevention, and Recovery

Springer

P. Andrew Karam
Mirion Technologies
Brooklyn, NY, USA

ISSN 1613-5113 ISSN 2363-9466 (electronic)
Advanced Sciences and Technologies for Security Applications
ISBN 978-3-030-69164-6 ISBN 978-3-030-69162-2 (eBook)
https://doi.org/10.1007/978-3-030-69162-2

This Springer imprint is published by the registered company Springer Nature Switzerland AG
The registered company address is: Gewerbestrasse 11, 6330 Cham, Switzerland

This book is dedicated to those who got me started and influenced me on this particular path of both service and science:

Fred Byus, Fellow Submariner (retired)
Audeen Fentiman, Ph.D., Doctoral advisor
Miguel (Mike) Frances, Sergeant, NYPD (retired)
Kevin Grogan, Detective, NYPD (retired)
Rupprecht Maushart, editor and mentor

Foreword

"With a screwdriver." That was how Robert Oppenheimer, the father of the American atomic bomb, answered the question of how a nuclear weapon smuggled into the USA in a crate could be detected. His answer, given in 1946 during congressional testimony, suggested a screwdriver would be necessary to open every crate and look inside. As some experts in the field are aware, Oppenheimer's answer led to the first US report on preventing nuclear terrorism, the so-called screwdriver report, which remains classified.

The number and type of tools we have developed to prevent nuclear terrorism have expanded greatly since 1946. Opening crates with a screwdriver to look inside has certain advantages, but is of course implausible on a large scale. To deal with this challenge, we have created an array of new radiation detector materials and devices that can help to detect illicit nuclear materials. A lightly trained police officer can now use a handheld detector to identify nuclear or radiological materials that only a trained scientist with a laboratory full of equipment could accomplish at the dawn of the nuclear era. Moreover, not far ahead, networked detectors using advanced anomaly detection algorithms will provide a detection capability many times more effective than the same number of individual detectors.

Of course, technology development is the only one element of an overall strategy to prevent nuclear terrorism, and notwithstanding important technical improvements in our capabilities, basic laws of physics will always work against detecting nuclear materials at a distance.

Technology has changed, but so has the risk. Nuclear and radiological materials are now used around the world in a variety of applications. In the early days, just a handful of countries produced nuclear weapons usable materials. The number of counties that possesses nuclear weapons usable material has grown substantially. Though the growth has not been as rapid as some feared, as reflected in JFK's 1963 prediction that in the 1970s "15 or 20 or 25 nations may have these [nuclear] weapons", the growth has continued. And even countries that do not seek nuclear weapons, such as Japan, possess substantial amounts of nuclear weapons usable material in civil stockpiles. Moreover, the technologies useful in the production of nuclear material and nuclear explosives have become much more easily accessible. Finally, with the dissolution of the Soviet Union in 1991, small, but still alarming amounts of material entered into illicit commerce for the first time. What will happen as governments in control of nuclear materials fall or become unstable in future?

Radiological materials are a separate problem—less threatening than nuclear materials, but with a much lower bar for acquisition. Radioactive materials are used widely in industrial and medical applications around the world—in all countries. Sometimes radioactive materials are used in quantities and forms that could be transformed into a significant radiological dispersal device—a so-called dirty bomb. Prior to 9/11, controls on these materials were weak. Since then, several countries, in particular the USA, have engaged in a coherent effort to help secure or replace these sources with non-radioactive alternatives, but the job is far from done.

As illustrated above, both the risk of nuclear or radiological terrorism as well as the technology to address this risk have changed substantially over the past seven decades since Oppenheimer made his screwdriver remark. However, one thing has not changed. And that is the need for practitioners and policy-makers to make decisions with a thorough understanding of the science that underlies nuclear science and technology.

That is why this book by Dr. Karam is so important. Along with a well-crafted curriculum, it can help to prepare the next generation of nuclear specialists in key areas related to nuclear terrorism. Scientists, technicians, politicians and policy-makers need a clear-eyed understanding of the science and the tools we must deploy to deal with these risks. Our ability to prevent a nuclear catastrophe will be based on the skills and knowledge of key decision-makers and implementers. This excellent book will help to facilitate that. It addresses a range of essential topics in the field.

Dr. Karam's work as a scientist and as scientific advisor for the Counterterrorism Division of the NYPD expanding and operating our rad/nuke interdiction system provides an excellent background for this book. He is a respected health physicist with experience in radiological and nuclear emergency preparedness and response. He has extensive teaching experience at all levels of education and has published widely for both technical and popular audiences. His professional skills have also been recognized by his colleagues; he was asked to participate on committees for the National Academy of Science, the National Council on Radiation Protection and Measurements, and participated in several overseas missions on behalf of the International Atomic Energy Agency and the Health Physics Society.

Seventy years into the nuclear era, no terrorist has successfully exploited the potential of nuclear power. Why is this? Is it because we have developed strong enough protections to prevent terrorists from succeeding? Is it because we have been successful in convincing terrorists that they will not succeed and thus they do not try? Is it because terrorists just are not interested or capable? We just do not know.

We do know that the threat will likely grow. And we know that technology alone cannot prevent nuclear terrorism. It must be integrated into an overall strategy, and the strategy must be adequately resourced. Preventing nuclear terrorism clearly requires a multi-prong effort. The first is to minimize the use of the most threatening materials. The second is to ensure that nuclear and radiological materials are secured at their source. The third is to ensure that we have in place the necessary technology and agreements to support recovery of nuclear or radiological material that may be stolen.

And finally, we need to be prepared to mitigate the effects of an attack through well-defined emergency response and consequence management plans and capabilities.

If we are lucky, we will have time to prepare. Do we have enough time? We may, but we do not have time to waste.

Warren Stern
Deputy Chair, Nonproliferation and National Security

Preface and Acknowledgements

I spent the first part of my adult life in the American Navy—8 years during the final decade of the Cold War, with half of that spent on a nuclear attack submarine. Those years were tense; protests against American nuclear weapons roiled Europe, the Soviet Union shot down a civilian airliner over the Kamchatka Peninsula, and Ronald Regan jokingly announced that he had outlawed the Soviet Union. The submarine I was on spent about two months every year in close proximity to the Soviet Union, following their submarines and collecting intelligence, and they sent ships and submarines to try to find and to follow us. During these years, too, I had the opportunity to make radiation measurements on and near nuclear weapons, giving me an appreciation for not only their destructive power, but also of how difficult detecting them can be. Our primary worry was that there would be a nuclear war between NATO and the Warsaw Pact that could quite possibly kill hundreds of millions of people.

Imagine my surprise when, shortly after I left the Navy, the Berlin Wall fell, followed in the next few years by the Warsaw Pact and the Soviet Union. The future seemed brighter than I could have imagined just a year or so earlier when we were trying to avoid detection by the Soviet Navy.

Fast-forwarding a decade, I have to admit to mild bemusement when some guy I'd sort of heard of—Osama bin Laden—"declared war" on the USA. But then his forces attacked two of our embassies in Africa, followed by a series of smaller attacks and then the attacks of September 11, 2001. And with the latter, my whole world changed.

As a radiation safety officer, my primary concern had been worries that an angry graduate student would dump some radioactive reagents into their supervisor's lunch—something that had happened a number of times at a variety of universities. And suddenly I was worried about terrorists trying to steal our high-activity irradiator sources to use in a terrorist attack—I had to learn about vehicle-borne explosives, how to work with police and fire to thwart or to respond to an attack, how to set up security systems, and any number of things I never expected to need to know about. As time went on, I found myself becoming increasingly immersed in a topic to which I'd never before had to give more than passing attention—I was asked to provide feedback on a proposed Congressional bill to help prevent radiological terrorism, I helped prepare our medical center to respond to a radiological or nuclear attack (or to provide assistance to others who were attacked), I contributed to a few reports on topics related to radiological and nuclear terrorism, and much more. And then things got even more involved.

I spent nearly a decade working for the New York City government, first for the Health Department and then for the Counterterrorism Division of the New York City Police Department (NYPD), getting two more views of this issue. My time at Health was spent working with my staff and with other agencies to develop the department's and the city's emergency response plans; at the Police Department, I continued this planning work as well as working with expanding and operating our rad/nuke interdiction system. During my time with Health and Police, I also spent a fair amount of time in the field, standing by with the Hazardous Materials Response Unit for major events, in the event that there was an attack, performing aerial and maritime interdiction surveys and helping to interpret the various instrument readings when I was with the Police. It is safe to say that the possibility of a radiological or nuclear attack changed the direction of my career and much of my life.

In the last 30 years, the world has grown both more and less dangerous and a lot more confusing. While we no longer worry that the world will be engulfed in nuclear fireballs, we do worry that a single city might; and while we are no longer faced with a relatively monolithic and implacable foe, we are facing a multitude of smaller enemies who, unfortunately, are not tied to a single state or even to a single location—an enemy that has few worries of being destroyed because we might not even know where to find them. Not only that, but our enemies' goals are different than those of the Soviet Union—they do not mind launching "nuisance" attacks with radiological weapons and they do not possess a nation or substantial resources that can be easily targeted for retaliatory attacks. Thus, we find ourselves in a world in which a nuclear attack is plausibly more likely than it was—but is less likely to be the existential threat it once was—and in which a radiological attack is more likely still. Preventing, planning for, or responding to such attacks is a complex matter that involves a sometimes-bewildering array of knowledge and skills; more than any single person can master.

Radiological and Nuclear Terrorism is aimed at a broad range of readers, including radiation safety professionals, emergency response planners, emergency responders, medical professionals, public health officials and the elected and appointed officials who will be managing our preparations for—and response to—any such attacks. My intention is to give each of these groups enough information to help them begin to consider how they would address their own piece of this issue, to provide high-quality references to help them dig deeper (should they wish) and sufficiently broad view of this matter to help them to understand how their own work falls into the overall "big picture" of both preparation and response. Hopefully, too, having a broad understanding of this matter will help to appreciate the problems faced by others and the work that these others are doing.

I have wanted to write a book like this one for over a decade. When I first started thinking about it, I was convinced I could do it well—and every year I learn more, I do more, and I realize how little I actually knew when this thought first crossed my mind. My goal—my hope—is that this will be a fairly comprehensive reference on a topic that is, and that, unfortunately, is likely to remain timely for some time to come. I hope that you will find it useful—at the same time, I hope you never need to use it.

Finally, there are far too many people who have helped me with this than I can properly thank. First and foremost is Anneliese Kersbergen of Springer; Anneliese gave me the opportunity to write this book, and she has exhibited unflagging patience, support and a sense of humor, all deployed as appropriate. Others who have helped with information, encouragement, reviews, support and moreover the years—with this book and in this area include (in no particular order other than alphabetical)—with my sincere apologies to those who I failed to mention due to CRS (a condition my wife has noticed and labeled—Can't Remember Stuff):

Armin Ansari
Laura Atwell
Narenta Bici
Daniel Blumenthal
Brook Bruddemeier
Robert Brown
John Buchanan
Eliot Calhoun
Gaetano Casarella
Kerry Collison
Andrew Couchman
John Crapo
Peter Dagostino
Salvatore DiPace
Miriam Edvardsen
Mary Roach
Harry Salters
Richard Schlueck
Keith Spero
Charles Ferguson
Benjamin Geitner
Mary Gill
Abel Gonzalez
Fred Harper
Jeffrey Hart
Daniel Hogan
Robert Hodges
Robert Ingram
Tomoaki Ishigaki
Jauharah Khudzari
Gladys Klemic
Ourania Kosti
Jodi Lieberman
Daniel Magee
Julien Spruyette
Sriram Srinivas
Warren Stern
Mark Teitler
Sigurdar Magnusson
Chris Maher
Rupprecht Maushart
Patrick McElroy
Carol McGowan
Michael Morehouse
Steven Musolino
Tami Nolting
Daniel O'Keefe
Holger Peters
Diva Puig
Cameron Reed
Richard Rhodes
Timothy Rice
Michael Riggio
Robert Ullrich
Yuichi Yokoyama

Brooklyn, NY, USA

P. Andrew Karam

Contents

Chapter 1
Introduction

In a 1954 interview described in author Bob Considine's biography of General Douglas MacArthur [1] the General was describing his plans for winning the Korean War. They included dropping "30–50 tactical atomic bombs" on North Korea's military and then contaminating the land "…from the Sea of Japan to the Yellow Sea—a belt of radioactive cobalt. It could have been spread from wagons, carts, trucks and planes. It is not an expensive material. It has an active life of between 60 and 120 years. For at least 60 years there could have been no land invasion of Korea from the North. The enemy could not have marched across that radiated collar I proposed to put across Korea's neck." While MacArthur obviously never implemented his plan, it certainly captures some of the essentials of both nuclear and radiological weapons—using the sheer destructive power of the former and the long-term "staying power" of the latter to destroy an enemy and to deny them use of an area of land for an extended period of time. Those considering using such weapons in radiological attacks are likely making the same type of assessment.

Something else that can be inferred from this interview is that MacArthur lacked a detailed understanding about the biological effects of radiation and the amount that would be required to sow such a large area with enough radioactive cobalt to make the area unsafe for several decades. This, too, reflects a mentality that has been carried forward for more than a half-century—that radiation is uniquely dangerous and that only a small amount is sufficient to cause grave danger. The fact that it is not—and that so few actually understand this fact—could well be the most damaging aspect of radiological terrorism, not to mention our response to radiological or nuclear accidents (e.g. Fukushima) and their aftermath. In fact, this lack of information—among members of the public, the medical community, politicians, emergency responders, public health officials, and many more—was among the factors prompting this book.

In Part I of *Radiological and Nuclear Terrorism* we will review some of the fundamental science of radiation; it's basic properties, how it affects our health, and how we detect it. This fundamental information will inform the rest of the book, helping the reader to understand the scientific context surrounding many of

P. A. Karam, *Radiological and Nuclear Terrorism*,
Advanced Sciences and Technologies for Security Applications,
https://doi.org/10.1007/978-3-030-69162-2_1

the questions that must be answered in trying to avert, respond to, or recover from a radiological or nuclear attack.

Part II will discuss radiological weapons—how they work, their possible health and physical effects, and how the use of radiological weapons might affect society. And in Part III we will explore the same aspects of an attack using nuclear weapons. Both of these will use the science presented in Part I to help the reader to understand the topic and, more importantly, to understand what the actual—not imagined or feared—impact of such an attack might be. In particular, if such an attack does occur it's important to understand what these weapons can—and cannot—do if we are to be able to plan and to respond appropriately.

From here, we move on to some more applied topics. Part IV uses what we have learned about radiation instruments and about these weapons to discuss how an attack might be stopped before it can even take place. This interdiction activity is what many call the "Left of Boom" phase (meaning the activities that take place on an imaginary timeline *before* something explodes). And in Part V we explore how to use the preceding information to help inform an effective response to a radiological or nuclear attack—one designed to provide for the public safety while also keeping emergency responders safe.

Finally, in Part VI we will use what has been discussed to help to determine how to recover from such an attack. And, in spite of fears to the contrary, there is no reason why such a recovery cannot be accomplished—Hiroshima and Nagasaki, after all, are thriving cities today, and there is no reason to believe that even nuclear terrorism would leave a modern city permanently uninhabitable. We can draw upon our experiences from the past, in conjunction with the fundamental science, to help plan recovery efforts from even something as major as an act of nuclear terrorism.

One thing to bear in mind is that there has never been a terrorist attack using radiological or nuclear weapons. There seems to be compelling evidence that terrorist groups have an interest in them and there have been any number of instances of attempted smuggling of radioactive or nuclear materials [2]—but there has never been an attack using them. But even if these weapons are never used by terrorists—or by governments—we can assume that there will be future accidents,small, moderate, or large in scale. The information found in this book is equally relevant for any radiological or nuclear event, deliberate or accidental.

On a personal note, one of our children used to be scared of the dark. In actuality, it wasn't the dark that was frightening—it was what might be lurking in the dark. Similarly, it seems reasonable to assume that it's the lack of good information about radiation and its effects that contributes to much of the fear surrounding it, and this fear makes it a good weapon for terrorizing the public. We can hope that, by helping to spread *good* information, we can help to lessen the fear so that we—as individuals and as a society—might give radiation the respect it deserves, but not unwarranted and inappropriate terror (Figs. 1.1 and 1.2).

Fig. 1.1 Hiroshima shortly after the nuclear attack (US Army photo)

Fig. 1.2 Hiroshima's "Atomic Dome" today. Author's photo

References

1. Considine R (1964) The long and illustrious career of General Douglas MacArthur. Fawcett Publications, New York City
2. Langewiesche W (2008) The Atomic Bazaar: dispatches from the underground world of nuclear trafficking. Farrar, Straus, and Giroux, New York City

Chapter 2
Types of Radiation and Their Properties

Radiation and radioactivity are physical phenomena that have been well-described by science. An understanding of their physical properties and an appreciation of the relevant science is essential if we are to understand how to evaluate the risks they pose, the measurements we make, recommended safety measures, and so forth. For this reason, we will start with a review of the fundamental science that underlies the field of radiation safety and its applications. In addition, the reader should consider that, in the midst of a radiological or nuclear emergency, it might not be necessary to develop precise calculations—quick estimate is often good enough, with more detailed calculations coming when the emergency phase is over. At times, a rough answer that comes quickly can be more useful than a precise answer that arrives hours or days later. Finally, please note that this is a quick review of the basic concepts and is not intended to be comprehensive or in-depth; those interested in a more in-depth understanding of the topic should consider an introductory textbook (e.g., [1, 2], among others).

2.1 What Radiation and Radioactivity Are

Radiation, to a physicist, is nothing more—or less—than the emission of energy by an object and the transmission of that energy to a different location. A toaster does this; a toaster emits thermal radiation that is transferred to a piece of bread. To most members of the general public, radiation is a dangerous and mysterious phenomenon that glows, mutates, and kills. What the public thinks of as "radiation" is more specifically referred to as "ionizing radiation"—radiation that carries with it enough energy to strip an electron from an atom to create a pair of ions where there had previously been an uncharged atom. It is this process of ionization that can initiate the chain of events that can lead to health problems. The threshold energy for causing an ionization is about 5 electron volts (eV); for photons this corresponds

P. A. Karam, *Radiological and Nuclear Terrorism*,
Advanced Sciences and Technologies for Security Applications,
https://doi.org/10.1007/978-3-030-69162-2_2

to a wavelength of about 250 nm, which is solidly in the ultraviolet (UV). Thus, any photon radiation that is more energetic (shorter wavelength) than UV will be ionizing, while visible light and longer-wavelength radiation (including radio frequencies) is non-ionizing.

In addition to high-energy photons, ionizing radiation also includes charged particles (alpha and beta radiation) and uncharged particles (neutrons). Each type of radiation has its own physical properties that affect the manner in which it interacts with matter and within our bodies and that determines the instrument(s) used for detection as well as our radiation safety practices. These are described in more detail below.

2.1.1 Alpha Radiation

Alpha particles are relatively massive charged particles given off by heavy and unstable atoms. Alpha radiation is very damaging to living cells because of its relatively high mass and electrical charge. These same factors cause alpha radiation to lose energy quickly in tissue, with the result that alpha radiation cannot penetrate more than several microns into tissue; less than the thickness of the dead cells of the epidermis. Thus, as long as alpha radiation is neither inhaled or ingested and as long as open cuts and scrapes are covered, alpha radiation is harmless. Internally, however, alpha radioactivity is highly dangerous; the Russian spy Alexander Litvenenko was killed by ingesting very small amounts of alpha-emitting polonium (Po-210)—about as much as a single grain of salt. While there is no need for protective equipment to avoid tissue damage, protective clothing will prevent skin contamination that might later transfer to food or to other objects. In addition, respiratory protection will reduce the risk of accidental inhalation and covering open cuts, scrapes, and other injuries will reduce the risk of absorption into the blood stream.

Table 2.1 summarizes some important information about alpha radiation.

2.1.2 Beta Radiation

Beta particles are electrons or positrons—small, light particles—given off by radioactive atoms. Any element can have nuclides that give off beta radiation. Beta radiation is less damaging to living cells than alpha radiation, but beta particles penetrate more deeply into human tissue—up to about 1 cm. Accordingly, external beta radiation cannot damage internal organs, but high levels of beta contamination can produce skin burns ("beta burns") unless decontaminated quickly. Heavy clothing (e.g. firefighters' turnout gear and heavy gloves) will protect the skin from most beta radiation, and workers should wear respiratory protection if beta-emitting radioactivity might be airborne.

While beta radiation is normally in the form of negatively charged electrons, some neutron-deficient radionuclides emit positrons—the antimatter equivalent of

Table 2.1 Information about alpha radiation

What will emit alpha radiation	• Nuclides with high atomic number and high atomic mass (e.g. Am-241, Ra-226, U-235, etc.) • Smoke detectors, gas lantern mantles, welding rods, some rocks, nuclear weapons, some radioactive sources
Detectors and instruments	• Geiger Mueller "pancake" probes, alpha probes
Health effects	• Some alpha emitters are also toxic heavy metals (e.g. uranium) • No health effects from external alpha radiation • **Can be very harmful if inhaled or ingested**
Protective equipment	• Respiratory protection if activity is airborne • Cover open cuts, scrapes, and other wounds, gloves
Levels of concern[a]	• Greater than 10,000 cpm on the ground • Greater than 1000 cpm on the bare skin • Airborne alpha activity
Other comments	• Some alpha emitters (e.g. Am-241, Ra-226) also emit gamma radiation • The former Russian spy Alexander Litvenenko was killed by adding trace amounts of alpha-emitting polonium to tea

[a]Note that readings in Tables 2.1, 2.2 and 2.3 are given in cpm (what the instrument reads) and not in dpm (the activity that is actually present) to simplify comparison with the instrument being used. The count rates provided here are for a "pancake" type GM detector at a distance of about 2 cm from the surface being surveyed. Other detectors will produce different readings

Table 2.2 Information about beta radiation

What will emit beta radiation	• Any element can have beta-emitting nuclides • Many laboratory nuclides (e.g. H-3, C-14, P-32, S-35) • Newer self-luminous watch dials and some self-illuminating exit signs
Detectors and instruments	• Geiger counters (e.g. GM pancake probe) • Beta scintillation probes • Liquid scintillation counter (in the laboratory)
Health effects	• External beta radiation cannot expose internal organs • May cause shallow burns (beta burns) if bare skin is contaminated • Can expose internal organs if inhaled or ingested
Protective equipment	• Anti-contamination clothing (including heavy gloves if possible) to reduce skin contamination • Respiratory protection (if airborne) • Cover open cuts, scrapes, or wounds
Levels of concern*	• Greater than 100,000 cpm on the ground • Greater than 10,000 cpm on the skin
Other comments	• Many beta-emitters also emit gamma radiation (e.g. I-131) • There have been several instances of laboratory workers trying to poison co-workers with beta-emitters such as P-32 • Positron emission will be accompanied by 511 keV gamma radiation from electron–positron annihilation

*See comment to Table 2.1

Table 2.3 Information about gamma radiation

What will emit gamma radiation	• Industrial and medical nuclides emit gamma radiation (e.g. F-18, Co-60, Se-75, I-131, Cs-137, Ir-192, etc.)
Detectors and instruments	• Sodium iodide-type detectors (e.g. PRD, RIID, sodium iodide detector, etc.) • Ion chamber (to measure dose rate)
Health effects	• External gamma radiation can expose entire body (including internal organs) • Exposure to high dose in a short time can cause skin burns • Prolonged exposure to eyes can cause cataracts
Protective equipment	• Anti-contamination clothing to reduce skin contamination • Respiratory protection (if airborne) • Cover open cuts, scrapes, or wounds
When to worry	• Gamma dose rates greater than 100 μR/hr (1 μGy/hr) indicates the likely presence of radioactive materials • Gamma dose rates greater than 2 mR/hr (20 μGy/hr) might require a boundary to exclude members of the public • Gamma dose rates greater than 100 mR/hr (1 mGy/hr) might require dosimetry to enter • Gamma dose rates greater than 1 R/hr (10 mGy/hr) should be entered only if necessary • Gamma dose rates greater than 100 R/hr (1 Gy/hr) can be dangerous
Other comments	• Many gamma-emitting nuclides also emit alpha or beta radiation • Only gamma radiation will penetrate many packages or containers—other forms of radiation may also be present when opening a package, container, or vehicle emitting gamma radiation

electrons. When positrons are emitted they will encounter electrons and the electron and positron will annihilate each other, emitting two gamma rays, each with an energy of 511 keV.

Table 2.2 summarizes some important information about beta radiation.

2.1.3 Gamma Radiation

Gamma rays are photons, similar to the photons emitted by electric lights; the primary difference is that, like x-rays, gamma rays possess sufficient energy to pass through objects. Since gamma radiation can pass through the entire body, it can expose internal organs to radiation dose. Protective clothing will protect against skin contamination, respiratory protection will protect against inhalation and ingestion, and good work practices (e.g. time, distance, and shielding) will help to reduce radiation exposure. Due to the higher energy of gamma radiation (compared to x-ray photons), gamma radiation cannot normally be shielded by using lead aprons as is done with the

Table 2.4 Properties of neutron radiation

What will emit neutron radiation	• Neutron sources (soil gauges, well-logging sources, AmBe, and Cf-252 neutron sources) • Operating nuclear reactors • Particle accelerators (research, medical) • Nuclear weapons (low level)
Detectors and instruments	Neutron detectors (Helium-3, BF3, and others)
Health effects	• External neutron radiation can expose the entire body (including internal organs) • Many neutron sources also emit alpha radiation
Protective equipment	• Anti-contamination clothing to reduce skin contamination • Respiratory protection (if airborne) • Cover open cuts, scrapes, or wounds
Levels of concern	• Any sustained neutron dose rate above natural background levels should be investigated
Other comments	• Background neutron levels will vary widely • Different instruments will have different background neutron levels • Neutron levels can increase somewhat when near large structures (called the "ship effect") on the water or on land

lower-energy x-ray radiation. Table 2.3 summarizes some important information about gamma radiation.

2.1.4 Neutron Radiation

Neutrons are subatomic particles that are given off during nuclear reactions such as nuclear fission or from radioactive sources comprised of beryllium mixed with an alpha-emitting radionuclide (e.g. Am-241 and beryllium, or AmBe). Neutrons are highly penetrating and will pass through most solid materials; they are best shielded by using hydrogen-rich materials (e.g. water, plastic) and neutron-absorbing elements such as boron. Bombarding materials with neutrons can cause them to become radioactive. Due to their higher mass, neutrons are much more biologically damaging than are beta particles or gamma rays; neutron absorption can also cause materials to become radioactive. Table 2.4 summarizes some important information about neutron radiation.

2.1.5 Radioactivity Units

Radioactivity is measured in terms of the number of radioactive decays (also referred to as "disintegrations") an amount of radioactive material (a radioactive source) undergoes each second. The international unit, the Becquerel (Bq), is the amount of a radioactive material that undergoes 1 decay in one second. As with other SI units, modifiers (k, M, m, and so forth) are used to reflect multiples or fractions of a Bq (e.g. 25,000 Bq = 25 kBq).

In addition to the SI unit, there are places in which the "traditional" unit of the Curie (Ci) remains in popular, if not in official, use. One Ci is that amount of a radioactive source that undergoes 37 billion decays in one second, so 1 Ci = 37 GBq.

It is important to note that the amount of radioactivity present in a source is not necessarily related to the physical size of the source. One gram of Ra-226, for example, contains about 1 Ci (37 GBq) of activity, one gram of Co-60 contains over 1000 times as much radioactivity, and one gram of U-238 contains less than a millionth as much radioactivity. Thus, 1 g of U-238 is radiologically harmless, 1 g of Ra-226 must be handled with caution, and 1 g of Co-60 can pose a risk to life and health with only several minutes of exposure. This is why the relative risk posed by a radioactive source can only be evaluated by making radiation measurements or through calculation (if sufficient information is available).

If one knows the half-life of a radionuclide and can calculate the number of radioactive atoms in a sample then it is possible to calculate the amount of radioactivity present in that sample using the law of radioactive decay: $A = \lambda N$ where A is the number of decays per unit of time and N is the number of atoms. The decay constant (λ) is the fraction of atoms that decay in a given amount of time and is equal to $\frac{\ln(2)}{t1/2}$ where $t_{1/2}$ is the nuclide's half-life. The amount of radioactivity per gram is referred to as the specific activity.

2.1.6 Radioactive Decay Calculations

When an atom of Co-60 has decayed, the progeny nuclide (Ni-60) is non-radioactive. This means that radioactive decay reduces the number of radioactive atoms and, as the number of radioactive atoms decreases so does the decay rate. In other words, the amount of radioactivity decreases with time. The magnitude of this decrease can be calculated fairly easily using the equation $A_t = A_0 \times e^{-\lambda t}$ where A_0 and A_t are the original and decayed activity (respectively), λ is the decay constant, and t is the time between A_0 and A_t. Decay can also be calculated using the number of elapsed half-lives: $A_t = A_0 \times 2^{-x}$ where x is the number of half-lives that have elapsed.

Example: calculating the specific activity of Co-60

- The number of atoms of Co-60 in one gram is 6.022×10^{23} atoms per mole/59.934 g of Co-60 per mole = 1.005×10^{22} atoms of Co-60 per gram.
- The decay constant for Co-60 = ln(2)/5.27 years = 0.1315 yr^{-1}.
- So the decay rate of 1 g of Co-60 = 1.005×10^{22} atoms of Co-60 × 0.1315 yr^{-1} = 1.32×10^{21} decays per year.

- Since there are about 3.156×10^7 s in a year, this comes out to about 4.2×10^{13} decays per second, or about 42 TBq.

There are a number of measurements whose change can be calculated in this same manner. Contamination levels, for example, are a measure of the amount of radioactivity per unit area (e.g. decays per second or decays per minute in a 100 cm^2 area); as the contaminating radionuclide decays, contamination levels will drop steadily. In a medical setting, for example, it is not uncommon to see contamination from Tc-99 m, a nuclide with a half-life of only about 6 h. With so short a half-life it often makes sense to simply lock the door to the room and let the contamination decay away over the space of a few days than it does to devote resources towards decontamination.

Example: calculating the decay of a Co-60 source

A radioactive materials licensee was cleaning out storage cabinets and found a 25-year-old Co-60 source with an original activity of 500 mCi (18.5 GBq). The licensee, not wanting to overpay for the source's disposal, needed to calculate the decayed activity of the source.

$$A_t = A_0 \times e^{-\lambda t}$$

We know from the previous example that the decay constant for Co-60 is $0.1315\ \mathrm{yr}^{-1}$ (from the previous text box) so the calculation is:

$$A_t = 18.5\,\mathrm{GBq} \times e^{-(0.1315 yr^{-1}) \times (25 yr)}$$

The decayed activity was calculated to be 0.691 GBq, or 691 MBq—considerably lower than the original activity and much less expensive to dispose of.

Similarly, as we will see in the next section, the radiation dose rate from a beta or gamma source depends on the amount of radioactivity present; as these source decay to lower activities the radiation dose rate drops as well. Thus, the radioactive decay equation can be used to calculate the reduction in radiation or contamination levels over time.

2.2 Radiation Dose and Dose Rate Determinations

When considering the health effects of radiation exposure the single most relevant fact is the amount of radiation dose a person, animal, or organ has received. Some radiation effects (e.g. skin burns) do not occur until a threshold dose has been received while the risk of developing other radiation effects (e.g. cancer) is proportional to the dose received. Because of this, it is important to have a good understanding of what radiation dose is and how the dose (and dose rate) change with changes in distance, shielding, and the amount of radioactivity present. It is equally important to understand the units in which dose and dose rate are measured and the different sources of exposure.

2.2.1 Definitions and Units of Dose and Dose Rate

One of the single most important concepts in radiation safety is dose or exposure. Dose rates are used as the basis for posting regulatory boundaries, for calculating stay times in a radiological area, for determining what actions to take in the presence of radioactivity, for reconstructing a radiation dose, and more. And the total dose that a person received is not only used to demonstrate compliance with regulatory limits, but is also the single most important factor in trying to predict—or to attribute—the health impact of radiation exposure. There are a number of concepts associated with radiation dose and dose rate, each of which will be discussed below.

Absorbed dose is simply a measure of the amount of energy deposited per unit of mass. The Roentgen, a measure of absorbed dose which measures the creation of electrical charge in dry air, is considered to be an obsolete unit and it not widely used. Units still in common use are the rad (primarily in the United States) and the Gray (the SI unit). It is important to note that the rad and the Gray are both measures of energy deposition in any absorber; it is appropriate to speak of absorbed dose in air, in water, or in tissue.

The Gray is defined as the deposition of 1 J of energy per kilogram of absorber. The rad is defined as the deposition of 100 ergs of energy per gram of absorber. Doing the appropriate unit conversions shows that 1 Gy = 100 rads.

2.2.2 Effective and Equivalent Dose

Measuring or calculating energy deposition is a good start, but we are primarily interested in how radiation exposure will affect the health of the person(s) exposed. Different types of radiation are more or less effective at causing genetic damage; beta and gamma radiation for example tend to cause point mutations and single-strand DNA breaks while alpha radiation can snap chromosomes or can cause multiple sites of damage as they traverse a cell's nucleus. Because of this, alpha radiation is as much as 20 times as damaging to cells as are beta or gamma radiation.

For this reason, weighting factors are assigned to the different types of radiation to calculate the amount of biological damage caused by exposure to the radiation—these are referred to as the Quality Factor (QF) or the Relative Biological Effectiveness

(RBE) and they range from 1 to 20. This is called the **equivalent dose** and it is measured in units of Sieverts (Sv) internationally and rem in the United States. Multiplying the absorbed dose by the RBE tells us how much biological damage—the risk of developing cancer in the future as well as the risk of developing short-term ailments such as bone marrow or lung damage—was caused by the radiation exposure. For example, exposure to enough alpha radiation (which has an RBE of 20) to deposit 1 J/kg would produce an absorbed dose of 1 Gy and an equivalent dose of 20 Sv.

There are times that ingested or inhaled radioactivity will travel preferentially to a single organ—I-131, for example, is absorbed primarily by the thyroid while uranium will concentrate in the kidneys and the bones. Different tissues also have different sensitivities to radiation and are more or less liable to develop cancers; if we are to determine the risk to a person from an intake of such a radionuclide we must understand better how a particular organ is affected by the radiation and the risk this poses to the person. For this reason, each of the body's major organs and organ systems have been assigned an organ weighting factor that is used to determine the **effective dose,** which can be used to determine the risk to the person exposed as though their entire body had been exposed. These organ weighting factors are shown in Table 2.5.

For example, a radiation exposure of 1 Sv to the red bone marrow (which has an organ weighting factor of 0.12) produces an effective dose of 0.12 Sv, or 120 mSv. These terms can also be combined. A person who inhales enough Rn-222 to deposit 0.1 J per kg of lung tissue would receive a *lung dose equivalent* of 2 Sv (0.1 Gy × 20) because the alpha radiation that Ra-226 emits has a relative biological effectiveness (RBE) of 20; and an *effective dose equivalent* of 0.24 Sv (2 Sv to the lung × 0.12 organ weighting factor). The effective dose equivalent (abbreviated EDE) can help us to determine the risk to the person exposed from the radiation to which they were exposed, even if the radiation or radioactivity is only administered to a small part of the body.

Table 2.5 Radiation weighting factors for various tissues [3]

Tissue	Weighting factor	Tissue	Weighting factor
Gonads	0.08	Esophagus	0.04
Red bone marrow	0.12	Thyroid	0.04
Colon	0.12	Skin	0.01
Lung	0.12	Bone surface	0.01
Stomach	0.12	Salivary glands	0.01
Breasts	0.12	Brain	0.01
Bladder	0.04	Rest of body	0.12
Liver	0.04	Total	1.00

2.2.3 *Deep and Shallow Dose*

Radiation is attenuated as it passes through tissue; some forms and energies of radiation are more attenuated than others. Beta radiation, for example, penetrates no more than 1 cm into tissue; alpha radiation penetrates only a few microns; while x-ray, neutron, and gamma radiation can penetrate through the entire body. Thus, a person exposed to very high levels of external beta radiation might receive enough exposure to give them skin burns while suffering no exposure at all to deeper tissues and internal organs. For this reason, it is useful to measure (or calculate) radiation exposure to the skin and shallow tissues as well as to deep tissues. These are called shallow (or skin) dose and deep dose respectively (radiation dose to the lens of the eye, which can cause cataracts, is usually tracked as well, but is not important for the purpose of this discussion).

2.2.4 *Calculating Radiation Exposure*

Radiation exposure can be measured directly, but there might be times when it may not be safe—or even possible—to make a direct measurement of radiation dose rates. For example, if the theft of a high-activity source is reported, emergency responders will not be able to make dose rate measurements on the source until it is located. However, it can be useful to determine how close a responder can safely approach and to distribute this information to the personnel who might be sent to try to interdict the source or to those who might respond to its use in an attack to help them stay safe as they work.

This process can also work in reverse; if the Radiological Health and Safety Officer can measure radiation dose rate at a known distance from a source, they will be able to determine how closely responders can approach without putting themselves at risk, and might also be able to determine the amount of radioactivity present based on the measured dose rate. For this reason, it is worth discussing some different methods for calculating radiation exposure and how it is affected by various factors. At the end of this section there will be an example problem demonstrating how all of these factors can fit together.

2.2.5 *Gamma Constant*

Every gamma-emitting radionuclide gives off gamma photons with a unique set of energies. Remembering that radiation dose (and dose rate) is a measure of energy deposition and that radioactivity is a measure of the rate at which radioactive atoms are decaying per unit of time, we can see that it is possible to add up the energies of the gamma radiation emitted by a source of a given activity and the number of gammas emitted per second and to use that information to calculate the radiation dose from that source. In fact, these calculations have been performed for a large number of gamma-emitting radionuclides—the answer to this calculation is called the gamma constant and it typically presented in units of Sv hr^{-1} for each MBq at a distance of 1 m.

For example, the gamma constant for Cs-137/Ba-137 m is 7.789×10^{-5} mSv hr^{-1} per MBq at a distance of 1 m. So the dose rate from a 37 MBq source would be 0.00288 Sv/h or 2.88 mSv/h at a distance of 1 m from the source.

2.2.6 *Allowable Limit for Intake (ALI)*

Just as one can calculate radiation exposure from external gamma radiation, so can one calculate the radiation exposure from radioactivity that is inhaled or ingested. Performing these calculations is much more complex as it requires adding in the energies from beta radiation and understanding the biokinetics of the chemical form of the nuclide(s) in question. This information has been used to develop the concept of the Allowable Limit for Intake (ALI); the ALI is the amount of inhaled or ingested radionuclide that will produce a radiation exposure of 50 mSv (5 rem) to the whole body or 500 mSv (50 rem) to the most-exposed internal organ. Thus, if the amount of intake is known, one can calculate the amount of radiation exposure received by comparing the intake to the ALI. An intake of 1 ALI will produce a dose of 50 mSv, a dose of 3 ALI will produce a dose of 150 mSv, a dose of 15 mSv will result from an intake of 0.3 ALI [7].

The greatest difficulty in calculating radiation exposure using the ALI is in quantifying the amount of intake. This can be done in the laboratory—analyzing urine or fecal samples, performing chest or whole-body counting, or other technique—or a rough assessment can be made in the field using a quick field assessment technique [5].

2.3 Calculating Radiation Attenuation

Calculating or measuring radiation exposure is a start, but it does not take into account how that radiation exposure is affected by changes in distance or when radiation shielding is put between the radiation source and a person. Understanding how distance and shielding affect radiation exposure can help to determine the length of time responders can safely work in a given area, how closely they can approach a radioactive source, and how to help protect them during response to a radiological or nuclear emergency. In addition, physicians and other medical responders can use these principles to help minimize their own radiation exposure when caring for heavily contaminated patients or those with embedded radioactive materials.

2.3.1 Attenuation Due to Distance

One can use the gamma constant to calculate radiation exposure at a distance of 1 m (or some other reference distance from a source), but it is not likely that any emergency responder will be spending a great deal of time at exactly that distance. Luckily, the manner in which distance affects gamma radiation exposure is well-understood and the calculations are not difficult.

Radiation dose rate drops as the inverse square of distance from a source. Thus, if one doubles their distance from a source the dose rate is reduced by a factor of 4 and tripling the distance reduces exposure by a factor of 9. Moving towards a source causes dose rate to increase in the same fashion.

So—when flying radiological interdiction missions the recommended altitude is about 100 m. However, when the author flew such missions over Manhattan's high-rise district it was necessary to fly at an altitude of about 500 m or higher to avoid flying into the sides of the skyscrapers. This was desirable from the standpoint of safety, but it reduced the ability to locate weak sources of radiation because the dose rate was reduced by a factor of 5^2 (25) at the higher altitude.

The inverse square law can also be used during an emergency response. A person standing one meter from a high-activity (37 TBq) source of Co-60 will expose a person to a fatal dose of radiation in about 40 min. Moving just one meter—one large step—away from the source extends that time to nearly three hours. A second large step increases the time even further, to about five or six hours.

Finally, the inverse square law can be used to determine the location of radiation safety boundaries. Say, for example, the Incident Commander wishes to establish a "Hot Zone" boundary at a radiation dose rate of 20 μGy/h, and is measuring a dose rate of 0.5 μGy/h at a distance of 250 m from the source. The ratio of these dose rates (20/0.5) is 40 so the Hot Zone boundary should be established at a distance that is closer by a factor of the square root of 40 (about 6.3). Dividing 250 by 6.3 gives a distance of just slightly less than 40 m—the Incident Commander should establish the boundary no closer than about 40 m from the source.

Using the concepts of the gamma constant and attenuation due to distance makes it possible to construct a table such as the one below, in which the gamma constant for each of three radionuclides can be used to calculate the dose rate at a distance of one meter with the inverse square law used to determine the dose rate at a variety of distances as shown in Table 2.6 [4].

These calculations use gamma constants of 1.14 R hr^{-1} for Co-60, 0.301 R hr^{-1} for Cs-137, and 0.424 R hr^{-1} for Ir-192, all at a distance of 1 m from the source. Units are in the American system because of the source from which the table was taken. To convert activity to the SI system, 37×10^9 Bq = 1 Ci.

Table 2.6 Distance to the 20 μGy hr^{-1} boundary for a variety of sources

	Distance (meters) to a dose rate of 20 μGy hr^{-1}[a]					
Nuclide	Source activity (Ci)					
	100	200	500	1000	2000	5000
Co-60	250	350	550	350	475	750
Cs-137	125	175	275	175	250	400
Ir-192	150	200	350	200	300	450

[a]In the US members of the general public are not permitted to have unrestricted access to areas in which they might receive a dose of 20 μGy in one hour. Thus, it is not uncommon to establish a boundary at this dose rate

2.3.2 *Attenuation Due to Shielding*

In addition, gamma radiation is absorbed by material; any material (including air) between a person and a source of gamma radiation will reduce radiation exposure somewhat, and some absorbers (e.g. lead) are more effective than are others (e.g. water). In an emergency it is easier to use distance to reduce radiation exposure than to install shielding, at the same time, it is often possible to use improvised shielding or to take advantage of existing objects (e.g. masonry walls or buildings, vehicles, and so forth). While there is a formal shielding equation, there might not be the time or ability to perform such calculations during the emergency phase of a radiological response. For this reason, it can be useful to use the concept of half-value and tenth-value layers (HVL and TVL). The half-value layer is the thickness of a given material that will reduce radiation exposure from a particular radionuclide by a factor of 2; the TVL reduces radiation exposure by a factor of 10.

When using HVL and TVL to determine attenuation it is important to understand that every combination of radionuclide and shielding material has a unique HVL and TVL. The high-energy gamma radiation from Co-60 is more penetrating than is the intermediate-energy Cs-137 gamma; it takes a greater thickness of lead to reduce exposure from Co-60 than from Cs-137 so these two nuclides will have different HVL and TVL values. Similarly, concrete is less dense and provides less shielding than lead; accordingly, the TVL for concrete is greater than is the TVL for lead. Example values for HVL and TVL for several important gamma-emitting radionuclides are provided in Table 2.7.

Calculating radiation dose on the far side of the shielding using HVL and TVL values is straight-forward; radiation exposure is reduced by a factor of 10 for each TVL (and by a factor of 2 for each HVL) through which the radiation passes. So, for example radiation passing through one HVL will be reduced by a factor of 2 while two HVLs will reduce radiation exposure by a factor of 2^2 (a factor of 4) and three HVLs will reduce radiation exposure by a factor of 2^3 (a factor of 8). Similarly, radiation passing through two TVLs will be reduced in intensity by a factor of 10^2; to only 1% of the unshielded dose rate.

Table 2.7 Half value layer (HVL) and tenth value layer (TVL) thicknesses for several radionuclides. Values calculated using the RadPro Calculator [6]

Nuclide	Water		Concrete		Lead	
	HVL (cm)	TVL (cm)	HVL (cm)	TVL (cm)	HVL (cm)	TVL (cm)
Co-60	26.4	64.7	11.8	28.1	1.8	4.5
Cs-137	19.9	54.3	8.8	23.9	0.7	2.2
Ir-192	24.9	49.9	9.2	21.2	0.3	1.1
Am-241	15.4	32.6	2.4	5.3	< 0.1	< 0.1

2.3.3 *Example*

Now let's put all of this together in an example.

Suppose you are in the role of a radiological Subject Matter Expert and you receive a report that some police performing routine patrols are measuring a radiation dose rate of 2 μGy hr^{-1} at a distance of about 50 m from a storage locker.

Using the inverse square law, you can tell the Incident Commander that she should establish a Hot Zone boundary at a dose rate of 20 μGy hr^{-1}—this should be at a distance of about 16 m from the storage locker.

$$50\,\mathrm{m}\sqrt{\frac{20}{2}} \sim 16\,\mathrm{m}$$

You further calculate that the dose rate at a distance of 1 m would be 5 mGy hr^{-1} (5000 μGy hr^{-1})

$$2\mu\frac{\mathrm{Gy}}{\mathrm{h}} \times \left(\frac{50\,\mathrm{m}}{1\,\mathrm{m}}\right)^2 = 5000\,\mu\,\mathrm{Gy\,hr}^{-1}$$

One of the responding police officers uses a RIID (radio-isotope identifier) and finds that the radionuclide in the storage area is Co-60.

Using a gamma constant of 3.045 $\times$ 10^{-04} mGy hr^{-1} for 1 MBq for Co-60, you calculate that the activity of the source (assuming it is unshielded) is about 16,400 MBq

$$\frac{5\,\mathrm{mGy/hr}}{\frac{3.045\times10^{-4}\,\mathrm{m\,Gy}}{\mathrm{hr}}\mathrm{per\,MBq}} = 16,400\,\mathrm{MBq}$$

You secure the area and obtain a warrant to search the storage area. You find that the highest dose rates appear to come from a box that is about 30 cm $\times$ 30 cm $\times$ 30 cm in size. The Bomb Squad is reluctant to handle the box because it might be rigged to explode, so you are asked to determine the worst case. You calculate that, if the box is made of lead, the highest-activity source it might contain is about 16,400 GBq or 16.4 TBq).

> The TVL for Co-60 gammas in lead is about 5 cm. Fifteen cm of lead is 3 TVL, so the dose rate will be reduced by a factor of 10^3. Thus, the apparent activity of 16,400 MBq (shielded) is equivalent of 16,400 GBq, or 16.4 TBq, assuming that the box in which the source is contained is made of solid lead.

You advise the Incident Commander that this is a high-activity source that, if removed from its shielding, can be very dangerous to emergency responders.

References

1. Gollnick D (2011) Basic radiation protection technology, 6th edn. Pacific Radiation Corporation, Altadena CA
2. Johnson T (2017) Introduction to health physics, 5th edn. McGraw Hill Education, New York City
3. International Commission on Radiological Protection (2007) Annals of the ICRP Vol. 37 Nos. 2–4. The 2007 recommendations of the international commission on radiological protection. Elsevier Ltd, Amsterdam
4. New York City (2010) Radiological response and recovery plan. New York City Office of Emergency Management
5. Korir G, Karam P (2018) A novel method for quick assessment of internal and external radiation exposure in the aftermath of a large radiological incident. Health Phys 115(2):235–251
6. Şahin D, Şahin MS, Kayrin K (2020) RadPro calculator online calculation utility. https://www.radprocalculator.com/default.aspx. Accessed 11 May 2020
7. US Environmental Protection Agency (1988) Federal Guidance Report No. 11: Limiting Values Of Radionuclide Intake And Air Concentration And Dose Conversion Factors For Inhalation, Submersion, And Ingestion. Government Printing Office, Washington DC

Chapter 3
Health Effects of Radiation

Radiation frightens many people and the primary reason for this fear is their anxiety about radiation's health effects. Popular culture accounts of radiation exposure tend to show dramatic injuries, burns, suffering, and death—all things that most people prefer to avoid. In addition, we read about the effects of in utero radiation exposure—causing birth defects for example—as well as long-term effects such as cancer. With all of this, it is hardly surprising that people tend to be frightened of radiation.

The bad news is that radiation can, indeed, cause many of the health effects that worry people so much. The good news is that it takes far more radiation exposure to cause these effects than most people believe.

3.1 Radiation Effects on the Cell

Radiation can damage cells by directly striking the DNA and causing damage such as single- or double-strand breaks or point mutations. It's more likely, however, that the radiation will strike molecules in the cytoplasm, interacting with them and forming free radicals (molecules that are very chemically active) that go on to cause DNA damage. Free radicals are generated by any of a number of mechanisms—mitochondria leak free radicals for example, metabolizing our food can create free radicals, and so forth. Not only that, but the DNA damage caused by any of these factors appears to be largely the same; in other words, we can't "look" at a point mutation and tell if it was caused by radiogenic or mitochondrial free radicals.

When radiation passes through a cell the effects can be non-existent, profound or anything in between. Sometimes a gamma ray will pass right through a cell without interacting at all and sometimes the free radicals produced will recombine or be scavenged before they can reach the DNA. If radiation (or the free radicals is produces) does interact with the DNA then either the DNA will be damaged or it won't. If the DNA is damaged, that damage may be beneficial, harmful, or neutral (neutral

P. A. Karam, *Radiological and Nuclear Terrorism*,
Advanced Sciences and Technologies for Security Applications,
https://doi.org/10.1007/978-3-030-69162-2_3

damage is damage that has no effect on the cell—it may be in non-coding part of the DNA, or to a gene that is not active in the particular cell that was damaged), and harmful damage may be either lethal or sublethal to the cell. At this point, the only DNA damage that concerns us is sublethal damage to the DNA in a part of the genome that may be harmful. Although lethal damage is bad for the cell, the damage does not get passed on to progeny cells, so a lethally damaged call cannot go on to cause cancer.

However, the possibilities do not stop here, because our cells have DNA damage repair mechanisms. Although these mechanisms are very effective, they are not perfect. This means that any bit of DNA damage may be repaired properly, may be repaired improperly, or might not be repaired at all. It is at this point that DNA damage may become a mutation—a mutation is what happens when damage to our DNA becomes "fixed" and is able to be passed on to the next generation of cells. As with DNA damage, mutations may be good, bad, or neutral, and the detrimental mutations may be lethal or sublethal. And, as before, it is only the sublethal damage that is of interest to us, and then, only if it can cause the cell to become cancerous (Fig. 3.1).

The preceding paragraphs are a quick summary of the various possibilities that might follow exposure of a cell to radiation. Part of the reason for this summary is for the sake of completeness, but it also helps to make an important point—radiation is a weak carcinogen. If we sum up all the possibilities above, there are over 20 pathways that can be followed. Of these, only one (sublethal damage that is misrepaired or unrepaired and causes a cell to become carcinogenic) have a chance of causing cancer. Radiation is a carcinogen, but not a very effective one—not compared to many of the chemicals we work with.

In the next few sections, we will look at the effects of both acute and chronic radiation exposure on the organism, instead of the individual cells. First, however, it is important to distinguish between acute and chronic radiation exposure and between deterministic and stochastic health effects.

Acute exposure occurs when an organism is exposed to a high dose of radiation in a short period of time. This is characteristic of radiation accidents. Acute radiation exposure causes deterministic health effects such as skin burns, vomiting, and so forth. Deterministic health effect occur when a person is exposed to more than a threshold dose of radiation, and the severity of the effects increase with increasing dose. So, for example, radiation burns from 5 Sv to the skin are more severe than from 3 Sv to the skin, and 1 Sv is below the threshold dose and will not cause radiation burns at all.

Chronic radiation exposure causes stochastic health effects; effects that are probabilistic and that might not have a threshold for induction. A person exposed to 1 Sv of radiation over a lifetime has a greater chance of developing radiogenic cancer than a person exposed to only 0.5 Sv—but if a cancer does ensue it will be neither more nor less severe than cancer caused by a lower or higher radiation exposure.

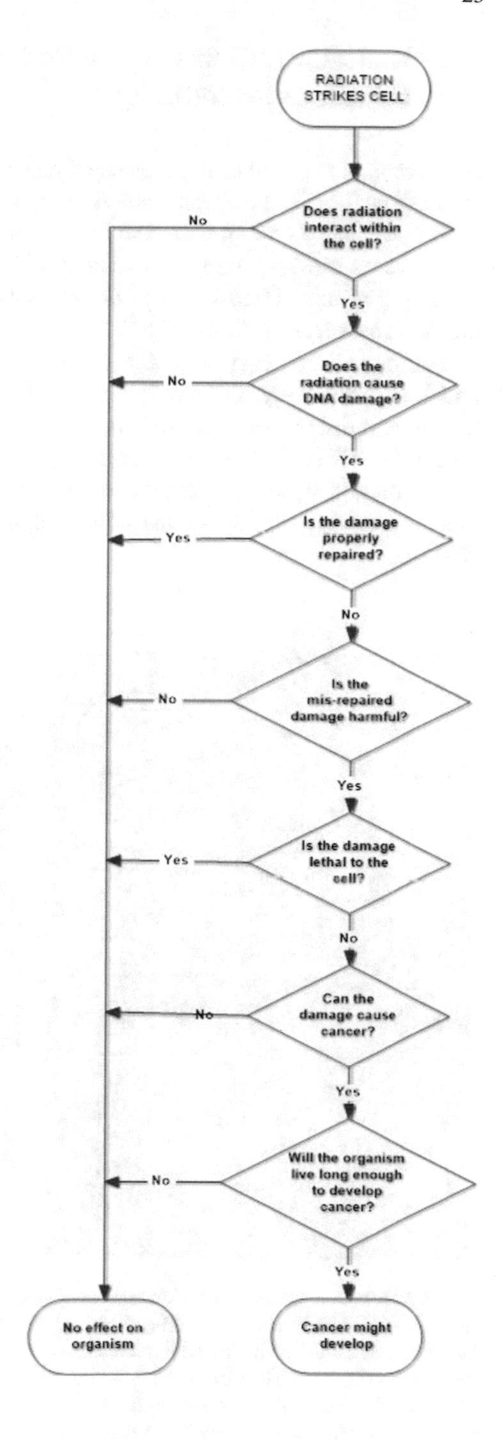

Fig. 3.1 Will DNA damage lead to cancer?

3.2 Radiation Effects on the Body Following Acute Exposure to Radiation

If a person is exposed to high levels of radiation in a short period of time, they will suffer from the effects of acute radiation exposure. If the dose is to a limited part of the bodies, they may end up with skin burns, sometimes severe. There have been many instances requiring skin grafts, excision of necrotic tissue, amputation of fingers, or even entire limbs. High levels of radiation exposure to the whole body can lead to radiation sickness or death (Fig. 3.2).

During radiological interdiction or response, workers might suffer radiation injury and it is most likely that any injuries will be radiation burns on their hands from handling radioactive sources or from putting their hands into a radiation beam. Those responding to a radiological or nuclear attack might find themselves confronted with high radiation dose rates from a radiation-emitting device (RED), a radiological dispersal device (RDD), or the fallout plume from an improvised nuclear device (IND).

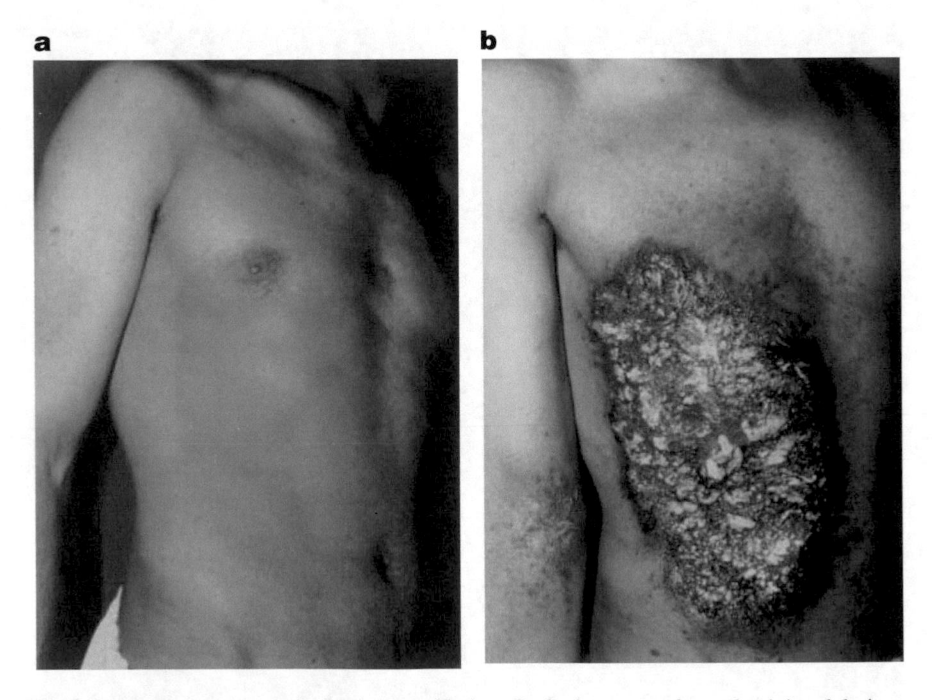

Fig. 3.2 **a**, **b** Changes to a radiation injury with time. Both photos are of a worker injured during a radiation accident in Gilan Iran. The worker placed a 185 GBq Ir-192 source into his shirt pocket for about 90 min. The photo on the left was taken 6 days after the exposure and we can see the reddened skin (erythema) from the radiation exposure. With time, the damage worsened; in the photo on the right the black area is necrotic tissue (radiation does not cause charring or thermal damage) and was taken 15 days after exposure. Used with the kind permission of the International Atomic Energy Agency [9]

The effects of high radiation dose to limited parts of the body may range from no observable effects (if the dose is low enough) to blistering, burns, or necrosis depending on the dose received. The effects of whole-body acute radiation exposure can be a bit more complex, and they are summarized in Table 3.1, derived from a variety of sources [1–3] and [10].

Table 3.1 Health effects of acute whole-body and localized radiation exposure

Acute whole-body dose (Sv)	Effect
0.01	Chromosomal changes (fragments, dicentric chromosomes, etc.)
0.25	Blood cell changes (depressed red and white cell counts)
1	Radiation sickness in about 10% of those exposed
4–5	Lethal dose to 50% of the population without medical treatment
7–8	Lethal dose to 50% of the population with medical treatment
10	Lethal dose to 100% of the exposed population
Acute localized dose (Sv)	*Effect*
2–6	Skin burns: 2–24 h post-exposure
2–5	Temporary epilation (hair loss): 2–3 weeks post-exposure
3–5	Dry desquamation (peeling): 3–6 weeks post-exposure
6	Permanent epilation: 3 weeks post-exposure
15–18	Wet desquamation (blistering): 4–6 weeks post-exposure
18	Necrosis: 10 weeks or more post-exposure

3.3 Radiation Effects on the Body Following Chronic Exposure to Radiation

The primary concerns with chronic exposure to relatively low levels of radiation are that we will develop cancer. There are two primary competing hypotheses regarding the risk of cancer induction from radiation exposure, the Linear, No-Threshold (LNT) model and Threshold models.

The LNT hypothesis suggests that *all* radiation exposure is potentially harmful ("no-threshold"), and that the risk of getting cancer from radiation is directly proportional to the dose received ("linear"). LNT is the most conservative radiation dose–response model in that it predicts the highest risk from a given amount of radiation exposure. This is one of the reasons that the LNT is the foundation of radiation regulations world-wide—since scientists are not certain as to how humans respond to low levels of radiation exposure, it makes sense to control dose (and risk) according to the most conservative model.

One concern with the LNT model is that it can be used to predict cancer risks down to vanishingly small levels of exposure, and so it has been used to calculate expected cancer rates from exposure to trivial levels of radiation. For example, say that the risk of getting cancer from a given radiation exposure is five additional cancer deaths for every 1 person-Sv. That means that exposing 100 people to 10 mSv each should result in an extra five cancer deaths among those people. Or, exposing one million people to 10 μSv each should also lead to five added cancer deaths. It's easy to see that we can use this model to predict added cancer deaths from any level of radiation exposure, no matter how trivial, if enough people are exposed. By analogy, we can also say that, since a 1000 kg rock will crush a person, throwing a million one-gram rocks at a million different people will cause one or more deaths by crushing. In reality, we know that this is not the case.

Because of this both the Health Physics Society and Roger Clarke, former head of the International Commission on Radiation Protection (ICRP) have advised against using the LNT model to predict risk at low levels of exposure. According to the Health Physics Society [4] it is not scientifically appropriate to calculate a numerical risk estimate from any exposure of less than 0.1 Sv. In a similar vein, the ICRP [5] has suggested that, when looking at the risk from collective dose, if the most highly exposed individual receives a trivial dose, then everyone's dose should be treated as trivial.

There are also studies that appear to show a threshold, below which no ill effects from radiation exposure are noted; still other studies seem to indicate a reduction in cancer from exposure to low levels of radiation—this latter effect is known as hormesis. While there are epidemiological studies that appear to support each of these models, the differences are not statistically significant, making it difficult to claim that any specific hypothesis is correct. Thus, regulations and risk estimates tend to be based on the LNT model as it predicts the highest risk for any given level of exposure. These three models are shown in very simplified form in Fig. 3.3.

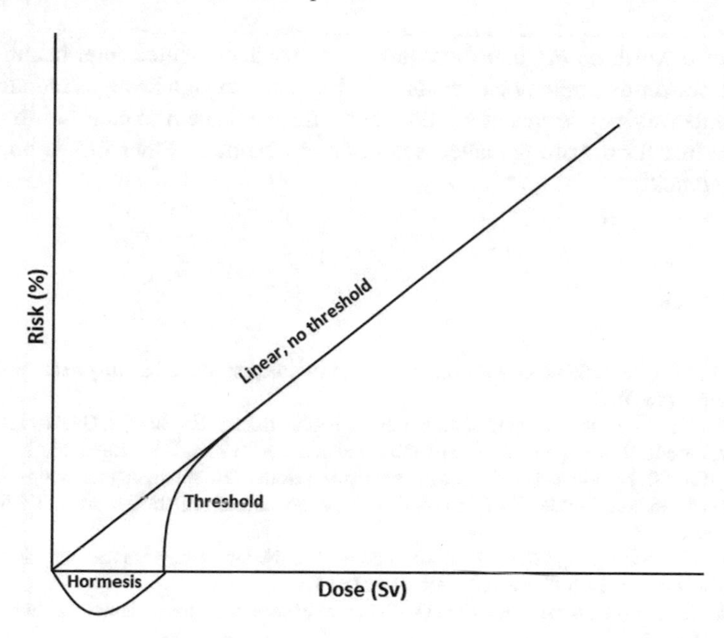

Fig. 3.3 Various radiation dose–response models; linear no threshold, threshold, and hormesis

3.4 Reproductive Effects of Radiation Exposure

Pregnant women are sometimes exposed to radiation occupationally or through medical procedures and, in the event of a radiological or nuclear attack there will undoubtedly be pregnant women exposed to radiation. Although even receiving a single x-ray can cause worry in an expectant mother, the amount of radiation required to cause harm to a developing fetus is higher than what most people (including many physicians) assume. The medical advice is to take no actions for any fetal radiation dose of less than 100 mGy and, in some cases, a dose of up to 150 mGy to the fetus poses little risk of causing birth defects or other problems [6]. Growth restriction is a possible outcome for a fetal dose of 100–500 mGy received between the third and thirteenth weeks post-conception, and higher doses can cause an increased risk of miscarriage as well as the possibility of neurological and motor deficiencies (CDC [8], NCRP [7]). This knowledge is the result of studying pregnant women receiving diagnostic or therapeutic medical radiation, Japanese women who were pregnant at the time of the atomic bombings in Japan, and pregnant radiation workers over more than a half century.

When advising a pregnant woman who has been exposed to ionizing radiation her physician should become familiar with the available literature and then request that a qualified medical physicist or health physicist calculate the fetal dose. The woman's physician will then have to weigh all the risk factors present and recommend a course of action to the woman.

It is also worth noting that there have been no documented fetal health effects from pre-conception radiation exposure [7]. Exposure to high levels of radiation may cause temporary or permanent sterility, but it does not seem to cause birth defects. Embryos that form from damaged sperm or ova seem to either fail to implant or miscarry quickly.

References

1. Hall E, Giaccia A (2006) Radiobiology for the radiologist, 6th edn. Lippincott Williams & Wilkins, New York
2. Mettler F (2001) Direct Effects of radiation on specific tissues. In: Gusev I, Guskova A, Mettler F (eds) Medical management of radiation accidents. CRC Press, Boca Raton FL
3. Sharp C (2001) Medical Accidents with local injury from use of medical fluoroscopy. In: Gusev IA, Guskova AK, Mettler FA (eds) Medical Management of radiation accidents. CRC Press, Boca Raton FL
4. Health Physics Society (2019) Position statement of the health physics society: radiation risk in perspective. Health Physics Society, Herndon VA
5. Clarke R (1999) Control of low-level radiation exposure: time for a change? J Radiol Protect 19(2):107–115
6. Wagner L, Lester R, Saldana R (1995) Exposure of the pregnant patient to diagnostic radiations: a guide to medical management, 2nd edn. Medical Physics Publishing, Madison Wisconsin
7. National Council on Radiation Protection and Measurements (2013) Report No. 174—Preconception and Prenatal radiation exposure: health effects and protective guidance. NCRP, Bethesda MD
8. Centers for Disease Control and Prevention (2020) Radiation and pregnancy: information for clinicians. https://www.cdc.gov/nceh/radiation/emergencies/pdf/303779-A_2019_Radiation-and-Pregnancy_508.pdf. Accessed 11 July 2020
9. International Atomic Energy Agency (2002) The radiological accident in Gilan. IAEA, Vienna
10. International Atomic Energy Agency (2020) Medical management of radiation injuries, Safety Reports Series No. 101. IAEA, Vienna

Chapter 4
Radiation Detection Technology

Although there have been anecdotal accounts of people sensing high levels of radiation (e.g. having a metallic taste or tingling skin) there is no scientific support for these accounts; to the best of our knowledge humans lack the ability to sense ionizing radiation. Thus, radiation detection instruments are the only reliable way that we have to indicate the presence of radiation or radioactivity and the level of risk that they pose. However, not all radiation detectors are capable of detecting all forms of radiation, conducting the different types of surveys that might be called for, or obtaining a reading under all conditions. As one example, radiation detectors made of sodium iodide are excellent for detecting gamma radiation, but are not at all sensitive to alpha particles; a policeman carrying a sodium iodide detector would be utterly unaware of the presence of even very high levels of Po-210, such as that used to assassinate Alexander Litvenenko. By the same token, the zinc sulfide crystals used to detect alpha radiation are not sensitive to gamma radiation. There are numerous examples of radiation instruments being used incorrectly or inappropriately, even by seasoned professionals.

Under some circumstances radiation can pose a threat to the health or the lives of emergency responders and/or members of the public. If those at the scene are using the wrong instruments then this risk might go unrecognized. For this reason, it is essential that those using instruments as well as those supervising their work have a good understanding of the strengths, weaknesses, and appropriate use for each of the instruments they purchase and use.

P. A. Karam, *Radiological and Nuclear Terrorism*,
Advanced Sciences and Technologies for Security Applications,
https://doi.org/10.1007/978-3-030-69162-2_4

4.1 Types of Radiation Surveys

There are three primary reasons for conducting radiation surveys:

1. To measure radiation dose rates. Radiation dose rate surveys are recorded in units of mGy/hr, μGy/hr (or their American counterparts) and they are used to:
   - Determine health and safety requirements,
   - Demonstrate compliance with regulatory requirements,
   - Identify the presence of unexpected radioactive materials.
     Note: Radiation dose is a measure of energy deposition; it requires the ability to measure the energy of the radiation detected by the radiation detector. Detectors that can register only count rates are not normally effective at accurately measuring radiation dose rate.
2. To measure levels of radioactive contamination. Contamination surveys are recorded in units of counts per minute (cpm) or counts per second (cps) and they are used to determine if protective anti-contamination clothing is required, if contamination levels are too high to permit an object, area, or person to be released from a contaminated area, or if decontamination is required.
3. To identify radionuclides and categorize them as natural, medical, industrial, and so forth

Note: some radionuclides can have multiple uses, and categorizing them simplistically can be misleading. For example, Cs-137 can be used for industrial radiography, temporarily implanted as a small source into tumors, or stolen for use in a terrorist attack; Ra-226 can be naturally occurring (present in granite or in sludge and pipe scales in the petroleum and natural gas industry), used in older compasses and watches, present as a contaminant in areas formerly used for the production of self-luminous products, or can be accidentally disposed of into the trash. In addition, the 186 keV gamma from Ra-226 is almost impossible to distinguish from the 185 keV gamma from U-235, the isotope of uranium used to make nuclear weapons. Although a radionuclide identifier might identify and categorize a radionuclide, those finding it must be able to understand the context in which it was found to understand the threat (if any) it poses (Fig. 4.1).

4.2 Gas-Filled Detectors

When ionizing radiation interacts with matter it will interact and strip an electron from an otherwise electrically neutral atom; this creates two ions, one positive and one negative. If this ion pair is created in a gas and the gas is embedded in an electrical field then the electrical field will cause the positive and negative charges to separate. If the voltage applied is high enough (a few to several hundred volts) the ions will be accelerated to high velocities. As they accelerate, they will interact with other atoms

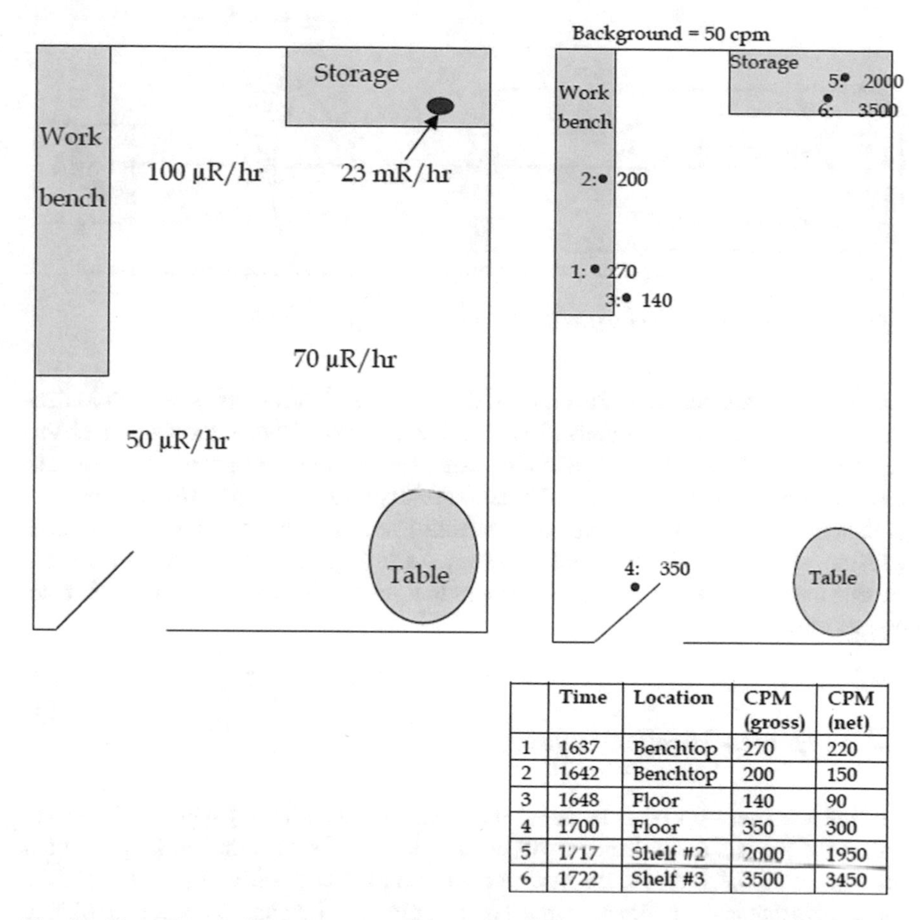

	Time	Location	CPM (gross)	CPM (net)
1	1637	Benchtop	270	220
2	1642	Benchtop	200	150
3	1648	Floor	140	90
4	1700	Floor	350	300
5	1717	Shelf #2	2000	1950
6	1722	Shelf #3	3500	3450

Fig. 4.1 Examples of radiation survey (left) and contamination survey (right)

in the gas, causing secondary ionizations. This process of gas amplification continues to build. At low voltages, the gas amplification is limited by the voltage (and, hence, the kinetic energy of each ion); at high voltages it is limited by the amount of gas present in the detector. Figure 4.2 shows what a basic gas-filled detector looks like.

4.2.1 *Ionization Chambers*

At low voltages—in what is called the "ionization region"—these changes in the electrical properties of the air inside the detector are proportional to the amount of energy deposited in the air. This means that we can directly measure energy deposition, which means that, at low voltages, gas-filled detectors can be used to accurately

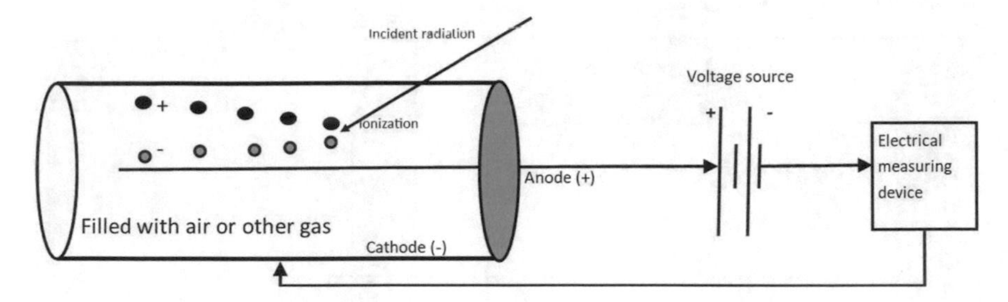

Fig. 4.2 Generalized sketch of gas-filled detector

measure radiation dose rate. Several varieties of ionization chambers (e.g. air ionization chambers, pressurized ionization chambers, tissue-equivalent ionization chambers) are ideal for making accurate radiation dose rate measurements. In addition to measuring gamma radiation, many ionization chambers are designed to measure beta radiation by using a plastic (typically bakelite) slab covering a thin metal window; with the beta shield open, beta radiation can pass through the window to be measured; closing the beta window screens out beta radiation so that only gamma radiation can be measured.

4.2.2 Geiger-Mueller Tubes

At high voltages—in what is called the Geiger-Mueller (GM) region—the entire volume of the gas-filled detector will become ionized through the gas amplification process. In the GM region, the instrument has maximal sensitivity to radiation; the trade-off is that every interaction in the detector looks the same, so it's not possible to determine the energy of the radiation that caused a count. This means that GM tubes are of only limited utility in measuring radiation dose rate, unless they are measuring radiation from the same radionuclide with which they were calibrated (e.g. Cs-137). Using a GM detector to measure radiation dose rate from Co-60 (for example) will produce a reading that is only about half as high as the actual reading—this can put people at risk.

For these reasons, GM detectors should not, as a general rule, be used to measure radiation dose rates unless:

- the radionuclide being measured is the same as the one with which the detector was calibrated, or
- the radionuclide is known and the user has a set of correction factors that can be applied to the instrument reading.

The exception to this is the energy-compensated GM, which is designed to produce accurate radiation dose rate readings across a wide range of gamma energies.

GM detectors are useful for performing contamination surveys (particularly a "pancake-type GM) and can be useful for measuring radiation dose rates if used as noted above. Some GM tubes have beta windows that can be exposed to measure beta radiation dose rate, but the dose rate readings suffer from the same sources of inaccuracy as noted for measuring dose from multiple energies of gamma radiation.

Geiger-Mueller detectors are relatively simple, inexpensive, and relatively rugged. However, the process of ionization, amplification, counting, and recombination requires a handful of microseconds and, during that time, additional counts will not be registered—this is called "dead time." Thus, GM detectors can saturate in high dose-rate radiation fields. If this happens with an older meter the needle might drop to zero because the detector is no longer registering individual counts; on newer (electronic) detectors the meter will typically indicate "overload" or something similar.

4.3 Scintillation Detectors (Handheld and Mobile Systems)

Gas-filled detectors are one major family of radiation detectors, scintillation detectors are the other. In a scintillation detector, ionizing radiation interacts with a material (usually a crystal or liquid) of some sort and causes it to emit photons of visible light; these photons can be collected and analyzed.

Unlike gas-filled detectors, which will react to any ionizing radiation that enters the detector, scintillation media are typically only used to detect and analyze one type of radiation; alpha, beta, or gamma. Sodium iodide and cesium iodide detectors, for example, are used to measure gamma radiation; zinc sulfide is used to measure alpha radiation; and various organic materials are used to measure beta radiation. Some of these compounds are sensitive to more than one type of radiation, but most scintillation media lack the generality of gas-filled detectors.

The properties of scintillation media are well-understood, including the number of scintillation photons produced for each MeV deposited in the detector. The energy of the scintillation photons is also well-known, as is the quantum efficiency of the photodiode and the amount of amplification of the signal as it progresses through the dynodes. Thus, it is possible to determine the amount of energy in the incident radiation by measuring the number of electrons collected from the instrument; this, in turn, can be used to identify the radionuclide that emitted the radiation.

The process of generating, amplifying, and processing each "count" takes a up to a microsecond [1] and any additional radiation that strikes the detector in that time will produce a pulse that overlaps with that of the earlier count—this is called "dead time." This can make it difficult to analyze each of these pulses and, if the radiation field is sufficiently high, it can become impossible to differentiate between counts; the detector can saturate, no longer providing useful information. Many detectors will identify when dead time becomes an issue making accurate nuclide ID and/or dose rate measurements difficult or impossible; other instruments will simply note "Overload" or the equivalent.

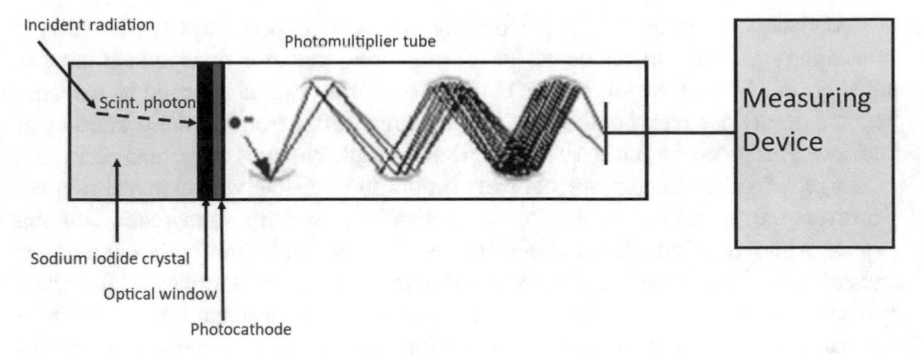

Fig. 4.3 Scintillation-type detector

Unlike gas-filled detectors, scintillation detectors can be relatively fragile and expensive and they can typically detect only one (or at most two) types of radiation (Fig. 4.3).

4.4 Neutron Detection

Neutron detection poses particular challenges as they are not detectable using common gas-filled or scintillation detectors [1]. With the exception of a material called "CLYC" (a crystal made of a cesium, lithium, yttrium, and chlorine compound) neutron detectors are only useful for detecting neutrons; outside of a nuclear reactor and neutron-emitting sources there is little need to survey for neutrons. Because neutron detectors are so specialized in their use and there are relatively few neutron sources, neutron detectors are far less common than those for other forms of radiation. In fact, if it were not for the fact that nuclear weapons emit neutron radiation there would be little reason for law enforcement departments to own them.

A common neutron detector is a gas-filled device using ^{3}He as the detection medium [2], owing to high detection efficiency of this isotope. However, as Kouzes notes, supplies of ^{3}He are growing sparse and increasingly expensive, prompting the development of alternatives such as CLYC, boron-trifluoride (BF_3), and detectors using lithium and other elements with a high neutron capture cross-section. Another neutron detection method is to use a small piece of highly enriched uranium and to detect the fission induced by the neutron field What these detectors all have in common is a low counting efficiency, making it difficult to perform interdiction surveys relying on neutron detection alone. However, even if neutron detection is difficult as a primary interdiction method, it is quite useful for confirming the presence of fissile or other neutron-emitting materials following initial identification through radio-isotopic identification (RIID), described next.

4.5 Radio-Isotopic Identification (RIID)

Because the number of scintillation photons produced is frequently proportional to the amount of energy deposited in the scintillation medium, this type of detector can be used to identify radionuclides.

Scintillation-type detectors are most commonly used for isotope identification because they are relatively inexpensive and operate at room temperature. However, scintillation detectors tend to lack good energy resolution and can fail to correctly identify closely spaced gammas. A classic example of this is ^{235}U versus ^{226}Ra, whose gamma energies are 185 and 186 keV respectively. To a scintillation detector, these energies cannot be distinguished; as a result, it is not uncommon for natural radioactivity to be identified as highly enriched uranium. This can also be used to mask threat nuclides by packing them in the same container as innocuous radionuclides whose gamma energies are close to those of the threat nuclide(s) and that are strong enough to hide the gammas from the threat (Fig. 4.4).

To some extent, performing a longer count can help to overcome these effects, by giving the detector (and the software) enough information that it can begin to tease out the weak signal of a masked radionuclide. Unfortunately, under routine circumstances it might not be possible to hold a vehicle or a person for an extended period of time. In addition, there are some gamma energies that are simply so close that even a long count with a scintillation detector will be unable to differentiate the energies.

Scintillation detectors can make mistakes

Another known property of scintillation detectors is that they can be sensitive to changes in temperature; the energy determination can change as the detector warms or cools. Newer instruments compensate for this by using a known gamma energy as a reference point to stabilize the instrument; older devices do not always do so.

As one example, the author was performing a radiological interdiction survey using a vehicle-mounted sodium iodide system in Midtown Manhattan on a cold day. At one point, the instruments began alarming with a ^{60}Co identification. Looking at the spectrum, it was clear that there was only one gamma peak (compared to the double peak of this nuclide), yet the "identification" continued to occur. The operator surmised that when the detectors had been initialized they were at the ambient temperature (about −2 °C) and that, as the vehicle warmed up, so did the detectors. As they warmed they began to "drift" until the 1.46 meV gamma from natural ^{40}K appeared to be the 1.33 meV gamma from ^{60}Co. To test this hypothesis, the instrument was turned off and then turned back on; it performed an automatic energy calibration at the new, higher temperature, and the ^{60}Co "identification" vanished. That particular instrument performed the start-up energy calibration using a small ^{137}Cs source; many newer instruments use natural ^{40}K and they continually monitor that gamma to stabilize the energy response during temperature changes.

Scintillation detectors can also misidentify radionuclides when two or more radionuclides emit gammas with very similar energies. As noted above, it is not uncommon, for example, to have Ra-226 (with a gamma energy of 186 keV) misidentified as U-235 (with a gamma energy of 185 keV).

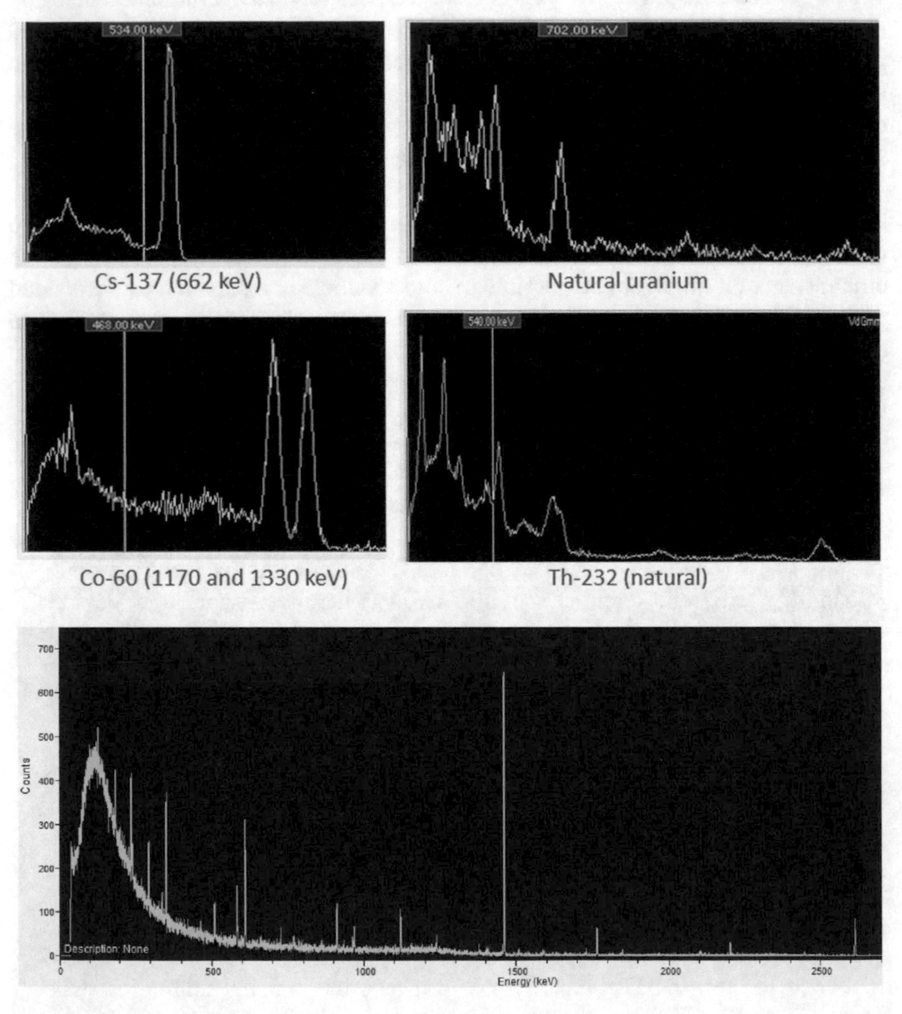

Fig. 4.4 Examples of gamma spectra, obtained with a sodium iodide detector (top four) and background radiation as measured with high-purity germanium (bottom). The high peak at 1460 keV is naturally occurring K-40. Spectra obtained by author

4.6 Instrument Selection

While only your radiation instruments will indicate the presence of radiological danger, excessive contamination, or other concerns, not every instrument is useful in all situations. A GM detector, for example, is not sensitive to low-energy beta and gamma radiation and does not display the correct dose rate in a mixed-energy gamma radiation field. Even radiation professionals sometimes select the wrong instrument or use the instrument they selected inappropriately. For this reason, it seems appropriate to briefly discuss how to select radiation instruments that are appropriate for the task at hand, and to summarize this information in a table.

4.6.1 Contamination surveys are best performed using a radiation detector with a large cross-sectional area that is sensitive to the type of radiation thought to be present.

- If multiple types of radiation are thought to be present, a "pancake" GM is useful, although this is not sensitive to low-energy beta (e.g. tritium) or gamma (e.g. ^{125}I) emitting radionuclides.
- Low-energy beta radiation can only be detected by performing a smear wipe that can be counted using a liquid scintillation counter or through the use of a hand-held proportional counter.
- Low-energy gamma contamination can only be detected by using a low-energy gamma scintillation detector, such as a thin-crystal sodium iodide detector.

4.6.2 Radiation dose rate surveys must be performed using a survey instrument sensitive to the type of radiation present and that is designed to accurately measure energy deposition (as opposed to simply recording counts).

Gamma radiation dose rate can be accurately measured using an ionization chamber or, at very low dose rates, using a pressurized ionization chamber. Gamma radiation dose rate can also be measured fairly accurately using an energy-compensated GM tube for gamma energies in excess of about 100 keV.

Beta radiation dose rate can be accurately measured using an ionization chamber with a "beta window."

Neutron radiation dose rate can only be accurately measured using a neutron detector.

Radiation dose rate from alpha radiation is not normally measured as, due to their very low penetrating ability, alpha particles typically do not exposure more than one or two cells at a time.

These are summarized in Table 4.1.

Table 4.1 Radiation instrument selection

Type of survey	Units	Detector	Comments
Contamination—alpha	cpm or cps	Zinc sulfide scintillation "Pancake" GM	Due to typically low counting efficiencies, alpha contamination surveys should be performed at very slow scanning speeds
Contamination (beta)	cpm or cps	"Pancake" GM Beta scintillator	A GM detector is sensitive to alpha, beta, and gamma radiation
Contamination (gamma)	cpm or cps	Sodium iodide scintillator "Pancake" GM	Sodium iodide detectors have a much higher detection efficiency for gamma detection Use a low-energy gamma detector for reliable counting of gammas with energy of <100 keV
Dose rate (beta)	μGy, mGy, or Gy hr^{-1}	Ionization chamber with beta window "Hot dog" GM with beta window	Some detectors require a correction factor to account for window size
Dose rate (gamma)	μGy, mGy, or Gy hr^{-1}	Ionization chamber Energy-compensated GM Sodium iodide detector	Sodium iodide detectors, as normally used, are frequently not energy-independent and should be used with the same precautions noted for GM detectors when used to measure dose rate
Neutron count rate	cpm or cps	He-3, BF3, Compensated ion chambers, CLYC[a], Lithium glass	Neutron count rate can be used for interdiction
Neutron dose rate	μSv, mSv, or Sv hr^{-1}	He-3 or other neutron detection media within a polyethylene moderator	Neutron dose rate is used for regulatory compliance and health and safety
Nuclide ID	N/A	Gamma scintillator (sodium iodide, cesium iodide, CZT, etc.) High-purity germanium	The user should consider the totality of circumstances when evaluating nuclide IDs that seem unlikely

[a]CLYC is Cs2LiYCl6:Ce (Cesium, Lithium, Yttrium, and Chlorine with Cerium impurities)

References

1. Knoll G (2010) Radiation detection and measurement, 4th edn. Wiley, New York
2. Kouzes R et al (2010) Neutron detection alternatives to ^{3}He for national security applications. Nucl Instrum Methods Phys Res A 623:1035–1045

Chapter 5
How Radiological Weapons Work

On May 8, 2002 Jose Padilla was arrested in Chicago under suspicion that Padilla was planning on constructing and setting off a radiological dispersal device (RDD)—colloquially known as a "dirty bomb." While Padilla was not charged with this crime, he was tried and found guilty of conspiring to commit murder and to fund terrorism in August, 2007—and both lawmakers and the public became aware of yet another threat [3].

That being said, these weapons were not a totally new threat. In actuality, the idea of radiological weapons goes back at least a half-century. During the Korean War, General Douglas MacArthur suggested sowing dangerous levels of radioactivity along the Korean-Chinese border to prevent further Chinese involvement in Korea following the presumed United Nation's victory in Korea [1]. Even earlier, in 1941, the National Academy of Sciences explored the idea of radiological warfare in the form of bombs that would distribute radioactivity in enemy territory [5]. In the 1980s, Saddam Hussein is thought to have experimented with Radiological Dispersal Devices (RDDs), eventually giving up on them for reasons to be outlined below [4]. And, in fact, the most recent incarnation of RDDs is not even a terrorist innovation; al Qaeda appears to have got this idea from watching US news broadcasts in the late 1990s [2].

In recent years, the first potential radiological terror event was threatened by suspected Chechen terrorists in Russia [17], who obtained radiological materials and set them out in public locations in Moscow as well as displaying them at a conference in the Chechen city of Shali. While these devices were never detonated, the fact that terrorists obtained radioactive materials and had the capability to use them in an attack brought this issue to the fore. It is also possible that al Qaeda produced an RDD [18], although if so, this weapon does not appear to have been used, and Ferguson (2019) mentions additional radiological plots, most of which never received a great deal of coverage in the media. In addition, since the early 2000s, Interpol and a variety of national and international security agencies have noted consistent attempts to traffic radioactive and nuclear materials [11, 13].

P. A. Karam, *Radiological and Nuclear Terrorism*,
Advanced Sciences and Technologies for Security Applications,
https://doi.org/10.1007/978-3-030-69162-2_5

5.1 Overview

At the most basic level, radiological weapons are fairly straight-forward—high-activity radioactive sources can be set out in such a way as to expose people to high levels of radiation, or they can be disseminated to spread radioactive contamination. Radiation exposure devices (REDs) would most likely be used to attempt to harm or kill people using radiation; radiological dispersal devices might be intended to try to cause harm, but are more likely to serve to spread non-lethal contamination to deny access to the contaminated areas and to cause economic and societal disruption due to the public's fears of radiation and radioactivity.

5.2 Types of Radiological Weapons

As noted above, the two primary categories of radiological weapons are those intended to irradiate people and those intended to spread contamination—radiological exposure devices (REDs) and radiological dispersal devices (RDDs) respectively. Consider, for example, a terrorist group that has stolen a high-activity (40 TBq) Co-60 radioactive source for malicious purposes. What impact can such a source have?

5.2.1 Radiological Exposure Device

The greatest health effects will come from keeping this source intact and concentrating the radioactivity to achieve the highest radiation dose rate. With a gamma constant of 3.945×10^2 mSv/hr per TBq at a distance of 1 m [12], this source would produce a radiation exposure of 15,780 mSv/hr, or slightly less than 16 Sv/hr at a distance of 1 m from the source. This will expose a person one meter from the source to a fatal dose of radiation in less than one hour, and any people within 2 m of the source will receive a fatal dose of radiation in 2 h or less (assuming they receive competent medical care).

What danger is posed by a 10 TBq source of Co-60 placed in a movie theater?
The gamma constant for Co-60 is 3.945×10^2 mSv/hr per TBq at a distance of 1 m, so the dose rate from this source would be about 3.9 Sv/hr at this distance—we will round this up to 4 Sv/hr to make the calculations simpler. If we assume that nobody leaves the theater, Table 5.1 show what the radiation doses to movie-goers will be at the end of a 2-h movie:

Table 5.1 Radiation exposure to movie-goers and the expected effects

Distance	Dose rate (Sv/hr)	Total dose (Sv)	Outcome	#People affected
30 cm	40	80	Death	1 (the person in the seat)
60 cm	10	20	Death	4 (both sides, front, back)
1 m	4	8	Death (50%)	8
1.5 m	2	4	Death (25%)	12
2 m	1	2	ARS	16–20

ARS Acute Radiation Sickness

So, for an attack of this sort, the maximum toll expected would be about one dozen deaths and two dozen victims of Acute Radiation Sickness—if everybody remains in their seats for the entire movie. In reality, it is likely that the theater would be evacuated when the first victims begin to vomit. Thus, the actual toll will likely be much lower than the numbers presented here.

The inherent limitation of an RED is the inverse square law—doubling the distance from the source reduces the dose rate by a factor of four. This fact, coupled with the fact that humans have a physical size and an aversion to spending long periods of time crammed into a small area, means that any RED is not likely to expose large numbers of people to a fatal dose of radiation—at least, not unless the radioactive source has a very high activity. But even a very high-activity radioactive source has limited utility as an RED. Consider, for example, a group that places the 40 TBq source inside a trash can on the street—the source will produce a fatal radiation dose in about 30 min at a distance of 1 m and a fatal dose in 2 h at 2 m distance. But how many people normally stand next to a mailbox for an hour or two at a time? People on the street typically walk or, at most, wait long enough for a traffic light to change. And even if people are standing in a group, there are only a limited number of people who will be standing within a meter of any location. In other words, human size and human behavior mitigate against this sort of a weapon being used to inflict fatal damage on more than a handful of people at any given time (Fig. 5.1).

Understanding this, a terrorist group might decide to try to plant a high-activity source in a location in which they expect people to remain relatively stationary for a prolonged period of time—at a sports arena or in a movie theater for example. Here, too, the inverse square law serves to mitigate the severity of the impact; even a long event (three hours) will not expose more than a few dozen people to a potentially fatal radiation dose, and then only if everybody remains in their seats for the entire event. With a 40 TBq source it might be possible to expose a dozen or so people to a fatal dose of radiation (again, assuming that they remain in place) and 100 or so might develop a survivable case of radiation sickness. But the overall toll is not likely to exceed that of all too many bombings we have seen in recent years. In other words, an RED is not likely to produce massive numbers of casualties.

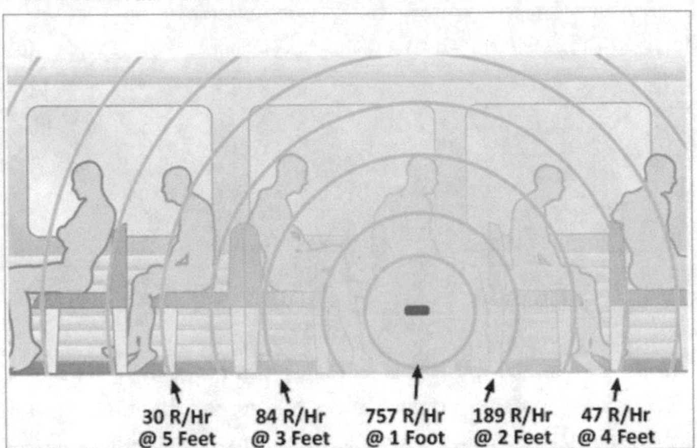

Fig. 5.1 Radiation-emitting device example. From US Health and Human Services (https://www.remm.nlm.gov/red.htm)

Another approach to an RED became known to the public in 2016, when the American Federal Bureau of Investigation (FBI) arrested mechanic Glendon Scott Crawford; Crawford was attempting to sell a "death ray" device that would be used to mow down Muslim on-lookers as a truck or van carrying the device drove past [21]. While the engineering and design details of the device Crawford was attempting to build have not been released to the public, it must be noted that here, too, it would be necessary to generate a beam of sufficiently high dose rate to produce a lethal dose of radiation in a short period of time, and that the inverse square law applies to this device as well. To produce a radiation exposure of 5 Gy (500 rad) in 5 s would require a dose rate of 1 Gy/s, or 3600 Gy/hr in the beam, and would require that a person remain in the beam at a distance of 1 m for 5 s. It is also worth noting that high dose rate x-ray beams are typically produced by linear accelerators that require several kilowatts of energy; such devices are large and they require a large power supply in order to produce dangerous radiation dose rates at any distance greater than a few meters. While such a device is not impossible, the logistics of its use suggest that this, too, is not an effective way to produce massive numbers of deaths from radiation exposure—especially not for a device mounted on a moving vehicle.

5.2.2 Radiological Dispersal Device

In the last section we saw what a radioactive source can do if it is used to cause the highest radiation dose possible; we should also explore the other extreme—what happens when the radioactivity is spread far and wide to contaminate the largest area possible. This would be accomplished by use of a radiological dispersal device

(RDD)—popularly known as a "dirty bomb," although there has also been speculation that disseminating radioactivity less obviously (a "smoky bomb") might also be a subject of concern [24].

5.2.2.1 Overt

An overt radiological attack is one that announces itself, classically as an explosion laced with radioactivity. In this sort of an RDD it is assumed that the radioactive materials would be placed on top of an explosive device and the force of the explosion would spread the radioactive materials—directly by the force of the explosion as well as smaller particles that would be carried by whatever wind or breeze might exist [7]. The materials carried by the force of the explosion would fly outwards on largely ballistic trajectories; the dispersal radius would depend on the force of the explosion, particle and fragment size, and related factors. For a large bomb (a few to several tons of TNT equivalent), fragments can be dispersed to a distance of a few to several hundred meters or more from the site of the explosion, carrying radioactivity to that distance. In addition, smaller particles (a few to several microns in size) will be lofted into the air and can be carried to a distance of several kilometers by the wind, breezes, and air currents. For a high-activity radioactive source (tens of TBq), radiation dose rates can be in the tens of mSv/hr over an area of a few hectares, while contamination carried by the breeze can form a plume that will deposit sufficient radioactivity to generate dose rates of up to a few tens of μSv/hr to a distance of three or four km downwind. Thus, it might be necessary to establish an extensive safety and security perimeter around the contaminated areas as well as those exposed to elevated levels of radiation by the radioactive source and source fragments.

The amount of contamination and its extent will depend on the source activity, the magnitude of the explosion, and the physical form of the source. Of these, the first two are self-explanatory but the last warrants discussion.

Liquids and powders are rather easily dispersible and will be relatively easily spread by a bomb. Some sources, however, are in the form of ceramic (e.g. Sr-90) or solid metal (e.g. Co-60, Ir-192)—such sources might flake or be broken into pieces (spalling) but are not likely to break into pieces small enough to be carried by the wind in the same manner as a powder or liquid. Thus, if solid sources are used in an RDD it seems reasonable to assume that radioactivity will not spread much beyond a few hundred meters, that the source fragments will contain higher levels of radioactivity (causing higher levels of radiation) than contamination from powdered or liquid sources, and that the resulting radiation dose rates will be somewhat patchy, with high dose rates in the vicinity of source fragments or shards and lower dose rates elsewhere.

Radiation exposure to members of the public will come from direct exposure to radiation from contamination and/or source fragments as well as exposure due to internal radioactivity from inhaled, ingested, or embedded radioactivity.

An overt attack is likely to cause injuries in the same manner as any other explosive device, with the added concern that these injuries may also be radioactively contaminated; for this reason it is reasonable to assume that injured persons more likely to have internal radioactivity than the uninjured. Even more importantly, those who are most seriously injured are likely to be the most heavily contaminated and to have the highest levels of internal radioactivity owing to their proximity to the explosion.

What size area can be affected by an RDD?

In the United States, a former regulatory guidance document [20] limited radioactive contamination in an uncontrolled area to 1000 dpm/100 cm^2 if the contamination is removable and 5000 dpm/100 cm^2 (50 dpm/cm^2) for fixed plus removable contamination. *So how large an area will be contaminated by a 10 GBq radioactive source?

$$10\,\text{GBq} = 10^{10}\,\text{dps} = 6 \times 10^{11}\,\text{dpm}$$

$$\frac{6 \times 10^{11}\text{dpm}}{50\text{dpm/cm}^2} = 1.2 \times 10^{10}\text{cm}^2$$

There are 10,000 cm^2 in one square meter, and 10,000 m^2 per ha, so the area that this source can contaminate to levels requiring regulatory access control would be 120 ha, or about 1.2 km^2.

It is worth noting that this level of contamination does not necessarily pose a risk to persons on the contaminated area. According to Federal Guidance Report #12 [22] Co-60 contamination produces a radiation dose rate of 2.35×10^{-15} Sv/s for contamination levels of 1 Bq/m^2. For simplicity, we can round off 50 dpm/cm^2 to be 1 dps/cm^2, which is 1 Bq/cm^2 because 1 Bq = 1 dps = 60 dpm. Thus, this level of contamination is about 1 Bq/cm^2 or 10,000 Bq/m^2 and will produce a radiation dose rate of 2.35×10^{-11} Sv/s. Bearing in mind that there are 3600 s in one hour, this is a dose rate of about 8.5×10^{-8} Sv/hr, or 0.085 μSv/hr. This is very similar to normal background radiation dose rates in most parts of the world and is not harmful.

*While this document was withdrawn in 2016, the contamination limits are well-entrenched in the US and are published in a number of US state regulations. They are used here for illustrative purposes.

5.2.2.2 Covert

In a December 19, 2006 New York Times opinion column, physicist Peter Zimmerman suggested that a terrorist attack involving radioactive materials might not be announced with an explosion. Using the term "smoky bomb" Zimmerman noted that covertly dispersing radioactive materials might have a greater impact on public health, especially if the radionuclide(s) used were the more-damaging and less-detectible alpha emitters such as the Po-210 used to assassinate former KGB agent Alexander Litvenenko in November, 2006 [24]. Zimmerman noted that, because it is so weakly penetrating, alpha radiation was not normally considered to be effective as an RDD isotope. But as the Litvenenko case demonstrated, relatively small amounts of alpha-emitting radionuclides, if ingested or inhaled, could be dangerous.

The most serious problem associated with a covert radiological attack is that nobody might know when the attack actually took place [10]. Consider a terrorist, for example, who surreptitiously spreads radioactivity in a heavily trafficked location such as a train station. Unless the station is monitored for radiation and/or radioactivity this will not be detected immediately, and possibly not for some time afterwards. In fact, the first detection might occur in a different location such as a university or nuclear power plant where surveys are performed on a regular basis. Now consider press reports that radioactivity was discovered in multiple locations around the city, spread by people passing through the train station, and the impact of this report on the city in question. As time goes on, more and more contaminated locations are identified and the citizens can only wonder how far it has spread, if they have been contaminated, and what the impact will be on their health and that of their friends and family.

This is the problem with a covert radiological attack—since nobody might know exactly when and where it occurred, the entire population of the city might consider themselves to be at risk, as well as commuters, tourists, and other visitors. This can obviously be disruptive to the city, businesses, and the people of the city.

However, launching a covert radiological attack is not necessarily a simple matter; especially in any of the many cities that are protected by robust interdiction networks, and even more so if the intent is to cause physical harm to those in the city. This is because gamma-emitting radionuclides must be shielded to prevent detection, and even alpha- and beta-emitting radionuclides frequently emit ancillary gamma or x-rays that can be detected at a distance (Am-241, for example, emits a gamma ray with an energy of about 60 keV, Ra-226 emits a 186 keV gamma, and so forth). In addition, beta-emitters produce x-ray radiation through the process of bremsstrahlung when the beta particles pass through shielding materials. This photon radiation can be detected, even if the particulate radiation is easily shielded.

As with an overt release, the covert dispersal of radioactivity is unlikely to pose a large threat to the health and safety of people living in the affected city. But even so, such an attack can have a significant financial and economic impact and can prove to be tremendously disruptive to the city and its citizens. Something else to keep in mind is that a covert radiological attack combines two significant fears—an instinctive fear of the unknown with the learned fear of radiation. Humans are instinctively frightened of what they do not know—the unknown can carry danger—and they learn to be frightened of radiation, partly because so few people have any real understanding of radiation and what it can do (i.e. fear of the unknown) and partly because what they do hear in the media is so often alarming or frightening.

5.3 Other Types of Radiological Attacks

There are other ways to launch radiological attacks. Some methods are speculative, others have been used already, albeit on a small scale—primarily adulterating food and/or drink with radioactivity (e.g. [23]). The purpose of this would likely not be to cause deaths so much as to frighten the public and cause them to lose confidence in the safety of their food and water.

There have been a number of deliberate attacks against individuals involving deliberate contamination and adding radioactivity to an individual's food or drink in an attempt to poison them (e.g. [15, 16]). Historically, this has happened primarily at research institutions, largely opportunistically when a researcher or graduate student becomes angry with a supervisor or co-worker. Part of the reason for this is that many researchers and graduate students have ready access to small quantities of radioactive materials in their laboratories and they are frequently under a great deal of pressure—if tempers flare or if they feel singled out for ill treatment it is easy to decide to slip some radioactivity into a person's food or drink.

Luckily for those attacked in this manner the amount of radioactivity found in most laboratories is not sufficient to cause radiation sickness or death; the person attacked must undergo bioassay monitoring, but there is little (if any) risk In addition, since these attacks are motivated by personal animus and not by a political agenda these attacks fall into the category of assault rather than terrorism. The case of Alexander Litvinenko, however, was different. Although this, too, was an attack against a single person, it was politically motivated—and it was fatal. The details of the Alexander Litvinenko assassination are fairly well-known; they are summarized in the accompanying text box and more details can be found in both the popular and the scientific literature (e.g. [8, 9]).

On a larger scale there is also concern that terrorist groups might seek to introduce radioactive materials into the food or water supply [24]. As noted above, this would very likely not cause harm or death to those exposed, but it could easily damage the public's confidence in their food and water. The National Council on Radiation Protection and Measurements discussed this possibility briefly [14], including several possible scenarios:

- Large-scale contamination of a municipal water supply
- Large-scale contamination of agricultural fields
- Large-scale contamination at one or more food processing facilities
- Small-scale contamination of food at the point of sale.

NCRP notes that an attack that deliberately targets the food and/or water supply can spread contamination from the transportation and handling of contaminated food and water and can expose individuals, possibly many individuals, to internal contamination from ingestion of radioactivity. NCRP also noted that attacks that take place further from the consumer—contamination of crops at a farm for example—will likely affect more people than contamination of individual items or individual sales locations (e.g. produce stands) closer to the consumer.

While there are certainly other possible scenarios for the misuse of radioactive materials by terrorist groups, anything more than the most cursory discussion brings one into areas that have been classified by many governments. Accordingly, they will not be further discussed here.

The Alexander Litvenenko assassination

On November 1, 2006 a former Russian spy named Alexander Litvinenko met two other Russians for tea at a restaurant in London. Later that evening Litvinenko started feeling ill and he was admitted to the hospital three days later. Three weeks later, on November 23, Litvinenko died; the following day, laboratory tests showed that Litvenenko had been exposed to Po-210, a radionuclide found in a handful of consumer and industrial products such as static eliminators.

Once the reason for Litvinenko's death was known a number of governmental agencies sprang into action. Law enforcement, of course, investigated the criminal aspects of the poisoning, concluding that Litvinenko had been poisoned when trace amounts of polonium were introduced to his tea at the restaurant, and ultimately determining that two Russians were the ones who brought the polonium into the UK and added it to the tea.

Public health officials were busy as well; in many ways they were even busier than the police. One public health task was to find out where the contamination had spread, which required trying to trace Litvinenko's footsteps around London, as well as those of his poisoners, both before and after the poisoning. Since Litvinenko's wife was likely contaminated, they also had to see where she had gone in the city. When all was said and done, over a dozen locations in London were found to be contaminated, as were three airliners, Emirates Stadium (in London), the British Embassy in Moscow, and a number of people that included staff at the hospital where Litvinenko was treated and restaurant workers who had served him and the Russians who poisoned him. A further 33,000 passengers had flown to over 50 different nations, requiring follow-up with public health officials in their home countries to check them for contamination. Laboratory and public health staff were also engaged in determining radiation exposure to all who were exposed and, had any of them had a significant intake of polonium, public health officials would have overseen the decorporation process.

Ultimately, the Litvinenko case involved law enforcement, intelligence, diplomatic, and public health officials in dozens of nations—in addition to the healthcare, food service, and airline workers and tens of thousands of members of the public. And all of this was from an amount of polonium that weighed about as much as a single grain of salt.

References

1. Considine R (1964) The long and illustrious career of general Douglas MacArthur. Fawcett Publications, New York City
2. Eng RR (2002) Medical countermeasures: planning against radiological and nuclear threats. Health Phys Soc Midyear Meet, San Antonio TX, USA
3. Fagenson Z (2014, Sept 9) U.S. Judge re-sentences Jose Padilla to 21 years on terrorism charges. Reuters
4. Ferguson CD. (2019) Past Illegal Attempts to Obtain Radiological Weapons. Presentation to Radiological Security: Scientific Approaches to Risk Assessment and Practicable Solutions. Russia-USEurope Workshop, Moscow Russia. December 3 2019
5. Federation of American Scientists (1998) Radiological weapons. https://fas.org/nuke/guide/iraq/other/radiological.htm. Accessed 8 Nov 2020
6. Ford J (1998) Radiological dispersal devices: assessing the transnational threat. Strateg Forum 136:1–3
7. Harper F, Musolino S (2007) Realistic radiological dispersal devices hazard boundaries and ramifications for early consequence management decisions. Health Phys 93(1):1–16
8. Harrison J, Fell T, Leggett R, Lloyd D, Puncher M, Youngman M (2017) The polonium-210 poisoning of Mr. Alexander Litvinenko. J Radiol Prot 37:1(266–280)
9. Harrison J, Smith T, Fell T, Ham G, Haylock R, Hodgson A, Etherington G (2017) Collateral contamination concomitant to the polonium-210 poisoning of Mr. Alexander Litvinenko. J Radiol Prot 37(4):837–851
10. Institute of Medicine Committee on R&D Needs for Improving Civilian Medical Response to Chemical and Biological Terrorism Incidents (1999) Recognizing covert exposure in a population in chemical and biological terrorism: research and development to improve civilian medical response. National Academies Press (US), Washington (DC)
11. International Atomic Energy Agency (2006) Combating illicit trafficking in nuclear and other radioactive material. IAEA nuclear security series No. 6. International Atomic Energy Agency Vienna 2007
12. International Commission on Radiation Protection (2008) Nuclear decay data for dosimetric calculations. ICRP Publication 107. Ann ICRP 38(3)
13. Langewische W (2008) The atomic bazaar: dispatches from the underground of nuclear trafficking. Farrar, Straus and Giroux, New York City
14. National Council on Radiation Protection and Measurements (2009) Report No. 161, Management of persons contaminated with radioactivity, vol 1. NCRP, Bethesda MD
15. New York Times (1982, Feb 24) Contamination of two is called deliberate
16. New York Times (1995, Oct 11) Radioactive poisoning alleged
17. NOVA (2003) Dirty Bomb Transcript available online at https://www.pbs.org/wgbh/nova/transcripts/3007_dirtybom.html. Accessed 9 Nov 2020
18. Salama S, Hansell L (2005) Does intent equal capability? Al-Qaeda and weapons of mass destruction. Nonproliferation Rev 12:3(615:653)
19. US Atomic Energy Commission (1974) Regulatory guide 1.86, termination of operating licenses for nuclear reactors
20. US Department of Justice (2016, Dec 19) New York man sentenced to 30 years for plot to kill Muslims
21. US Environmental Protection Agency (1993) Federal guidance Report No. 12: external exposure to radionuclides in air, water, and soil. Governmental Printing Office, Washington DC

22. Williams M, Armstrong L, Sizemore D (2020) Biologic, chemical, and radiation terrorism review. In: StatPearls [Internet] Treasure Island (FL): StatPearls Publishing. Accessed 10 Nov 2020
23. World Health Organization (2002) Terrorist threats to food: guidance for establishing and strengthening prevention and response systems. World Health Organization, Geneva
24. Zimmerman P (2006, Dec 19) The smoky bomb threat. New York Times

Chapter 6
Health Effects of Radiological Weapons

As noted in Chap. 3, one of the greatest radiation-related fears among the public is the fear that it will affect their health. People worry about radiation sickness, radiation burns, and death in the short term; they also worry about birth defects and cancer in the long term. Because the great majority of people do not understand the actual health effects associated with radiation exposure, there can be a tendency to attribute to radiation a greater deadliness than is actually the case.

In addition to the public's concerns, fears of radiation also affect emergency and medical responders and their readiness to go to work in the aftermath of a radiological incident. In a survey conducted in 2003, Martens and others reported that as few as 22% of emergency and medical responders stated that they were likely to report to work due to concerns about the impact of radiation and radioactivity on their health and that of their family members [5], a finding that is not uncommon [1].

6.1 Modes of Exposure

The fundamental modes of radiation exposure are internal and external: Internal radiation exposure from the inhalation or ingestion of radioactivity disseminated into the environment, and external exposure from a source set out as an RED or from radioactivity from an RDD that remains outside the body on the ground or suspended in the air. Radiation exposure can also come from pieces of radioactive material that become embedded in the body or contamination that is absorbed into the body through open cuts or scrapes or through mucus membranes.

External radiation exposure during an emergency will expose emergency responders and members of the public to relatively high doses of radiation in a relatively short period of time. If this acute radiation exposure is sufficiently high it can cause deterministic effects such as radiation sickness, skin burns, blood cell depletion, or even death. External radiation exposure can also come from radionuclides suspended

P. A. Karam, *Radiological and Nuclear Terrorism*,
Advanced Sciences and Technologies for Security Applications,
https://doi.org/10.1007/978-3-030-69162-2_6

in the air, although it will last only as long as the radioactivity remains suspended; as it settles out the radioactivity will collect on the ground where it can continue to emit radiation until it is remediated or decays to stability. In addition, very high levels of ground contamination can become resuspended by the wind or the passage of vehicles, becoming an inhalation hazard.

Over a longer period of time, internal radioactivity can become incorporated into the body, some of which might remain for the rest of a person's life. Radium, plutonium, and americium (for example) will become incorporated into the bone where they will remain for decades, leaching out slowly over the years. Iodine and cesium, by comparison, leave the body fairly quickly and expose the body to lower levels of radiation since the exposure ends more quickly than with more persistent radionuclides. This chronic radiation exposure will increase the person's risk of developing cancer during the remainder of their life, although the risk might be low.

6.2 Exposed Populations

Emergency responders are likely to receive acute exposure from their activities near high-activity sources and/or working in highly contaminated areas. In addition, any radioactivity that is inhaled or ingested during the course of their activities will continue to expose them to radiation for as long as the radioactivity remains in their bodies—in the case of bone-seeking elements with long half-lives (e.g. plutonium, radium, strontium) this chronic exposure will continue for the rest of the responder's life. Thus, responders who receive a high radiation exposure in a short period of time might suffer from acute radiation sickness, skin burns, and other acute effects of radiation exposure in the weeks and months following a radiological attack, with a continuing elevated risk of developing cancer in subsequent decades.

Members of the general public are also likely to receive acute exposure in the same manner as are emergency responders, although their radiation dose might well be lower if they are distant from the source(s), or if they are quickly evacuated through the areas with the highest radiation levels. However, members of the public who are in the immediate vicinity of a release of radioactive materials might well be more likely to experience an intake of radioactivity due to their lack of protective equipment—unlike emergency responders most members of the public are likely to lack respiratory protection, anti-contamination clothing, to undergo proper decontamination (including monitoring themselves to evaluate the success of their decontamination efforts), and those in the immediate vicinity of an explosive release of radioactivity might also suffer embedded radioactive fragments or contamination entering their bodies through open wounds. Thus, members of the public might also be expected to evidence both acute and chronic health effects from their exposure.

Radiation exposure to emergency responders

Let us assume that a terrorist group has set off a radiological dispersal device consisting of 1000 Ci (37 TBq) of Co-60. The force of the explosion distributes the majority of the radioactivity throughout an area about 100 m in radius, contaminating an area of about 7.8 acres (3.1 ha). The Incident Commander and/or Health and Safety Officer might be interested in determining how long emergency response personnel can safely work on this site.

For the sake of simplicity we will assume that the contamination is spread evenly across this site; spreading 37×10^{12} Bq across about 31,000 m^2 produces an activity concentration of 1.2×10^9 Bq per square meter. Federal Guidance Report #15 [10] notes a dose conversion factor for Co-60 distributed as contamination on an infinite plane will produce a radiation dose rate of 1.54×10^{-15} Sv s^{-1} Bq^{-1} m^2. Multiplying these two terms shows us that this level of contamination will produce a radiation dose rate of about 1.85×10^{-6} Sv s^{-1} or 6.7×10^{-3} Sv hr^{-1} (6.7 mSv hr^{-1} or 670 mrem hr^{-1} in American units).

In the United State emergency responders have a dose limit of 0.25 Sv in order to save property and a recommendation that individuals be limited to twice this dose (0.50 Sv) to save lives. This means that, if necessary, a firefighter could be permitted to work in this area for as long as 37 h fighting fires to save structures or could be engaged in life-saving activities for slightly more than three days. Note that these times refer to times working in the contaminated area, which means that each firefighter could make repeated entries over the course of several days or longer (assuming that they would exit the area from time to time to change air supplies, rest, eat, rehydrate, and so forth).

Considering that the minimum radiation dose needed to cause radiation sickness is about 1 Sv (100 rem) it seems unlikely that any firefighter in this scenario would be exposed to enough radiation to cause illness. Over a longer term, assuming a risk coefficient of 5% per Sv of exposure, these personnel would have an increased lifetime cancer mortality risk of no more than 2.5%. For members of the public who shelter in a stable and safe location until the emergency phase ends the dose (and resulting risk) would be significantly lower than to firefighters.

However, those members of the public who are not proximate to the release of radioactivity and those who are indoors (and who remain indoors until the dust has settled) are likely to receive little radiation exposure with the exception of that which they receive during evacuation from the affected area.

6.3 Special Considerations for Those Near an Explosion

Explosions are dangerous—this should go without saying. Explosions injure and kill people through the pressure wave they produce, by propelling shrapnel and other objects at high speed, by causing structures to collapse, and so forth. In addition to these a "dirty bomb" can also expose people to radiation, radioactivity blasted into the air can be inhaled or ingested, radioactive fragments or pieces of radioactive sources can be driven into flesh, and radioactive contamination can enter open cuts or wounds.

6.3.1 Ingested or Inhaled Radioactive Materials

Inhaling or ingesting radioactivity can result in a high radiation dose because, when radioactive materials enters the body, exposure continue as long as the radioactivity is present. All else being equal, inhaled radioactivity tends to produce a higher dose than when the material is ingested because materials in the digestive tract are typically present for only one or two days before the materials are eliminated and anything not absorbed into the body during this time will be excreted. On the other hand, materials in the lungs might remain there for weeks or months if the particles are inhaled deeply into the lungs—such particles might either dissolve totally into the body fluids and will continue irradiating the tissues of the lungs until they dissolve.

Once the radioactivity had entered the bloodstream it will circulate through the body according to its chemical properties. Strontium, radium, and plutonium (for example) will be taken up by the bones and will remain there for the remainder of the person's life, cesium will behave similarly to potassium and will pass out of the body relatively quickly, iodine will be taken up by the thyroid and will remain there for several weeks while iodine in the blood will be quickly excreted. Understanding the biokinetics of the radionuclide(s) inhaled or ingested is crucial to understanding the long-term radiation dose the person will receive [6].

For those who have ingested or inhaled radioactivity it will be important to quickly assess the approximate radiation dose to be expected; this can be quickly performed using hand-held radiation instruments (e.g. [3]), by making more precise (albeit more time-consuming) bioassay measurements [7], or by assessing a person's medical condition as their symptoms unfold [11]. If this assessment indicates that a person has ingested or inhaled a potentially damaging amount of radioactivity it is frequently possible to hasten the removal of this radioactivity from the body using an appropriate decorporation agent [7] to reduce radiation exposure. As one example, following the radiation accident in Goiania Brazil the use of potassium ferrocyanide (colloquially known as Prussian blue) was noted to reduce the biological residence time (and concomitant radiation dose) by as much as 43% [9]. Although there are not decorporation agents available for every known radionuclide, they are available for a large number of elements in various chemical forms.

6.3.2 Radioactive Embedded Objects and Contaminated Wounds

Embedded fragments of radioactive material might be as simple as micron-sized particles driven into the skin by the force of an explosion or they might be higher-activity pieces from a radioactive source shattered by the force of an explosion. The former offers little risk to the patient or to medical personnel caring for them; depending on the activity included in the latter, there might be a degree of risk to the patient or to those caring for them depending on the amount of radioactivity contained in the fragments. For example, a relatively low-activity source fragment containing 37 GBq (1 Ci) of radioactivity will produce a radiation dose rate of only about 10 mSv hr^{-1} (1 R hr^{-1}) at arm's length (about 1 m) and a dose rate as high as 1000 Sv hr^{-1} (10^5 R hr^{-1} to tissues in contact with the fragment. Dose rates this high can cause tissue damage in the patient and to the unwary surgeon who might handle the fragment when removing it from the patient.

In addition to radiation damage from embedded fragments, they can cause physical trauma as they pass through tissues; these fragments can also begin to dissolve into body fluids and enter the bloodstream, where the radioactivity can be carried to internal organs (where it can remain for days, months, even years) or to the kidneys to be excreted from the body.

Radioactive contamination can also enter the body through open cuts, wounds, and other injuries that can result from being in proximity to an explosion. Sweat, rain, water used for fire-fighting, even liquids used for decontamination can wash contamination into open cuts or wounds where the radionuclides can enter the bloodstream with the results noted earlier.

6.3.3 Blast-Related Injuries

Any medical responder—emergency medical technicians, paramedics, nurses, physicians, and others—must be able to prioritize the severity of various medical issues facing their patients so that they can properly perform triage. Many injuries—broken bones, lacerations, burns, and so forth—can be relatively easily seen and triaged in the field; explosives, however, have the potential to cause injuries that are not as easily seen. Thus, it seems reasonable to briefly discuss some blast-related injuries that might not be obvious during an initial examination. This is an overview only; for a more thorough discussion the reader should refer to the references noted in this section as an introduction to this subject.

Horrocks [2] describes four categories of blast injuries, Primary, Secondary, Tertiary, and Quaternary—each with its own set of characteristics.

Primary Interaction of the blast wave with the body. Gas-containing structures (ear, lungs, gastro-intestinal tract) are particularly at risk.

Secondary	Bomb fragments and other projectiles energized by the explosion cause penetrating and non-penetrating wounds. Any part of the body may be affected.
Tertiary	Displacement of the body (or of its constituent parts). This mechanism contributes to the traumatic amputation of limbs. Structural collapse of buildings (e.g. crush injuries).
Quarternary	A miscellaneous collection of all other mechanisms such as: Flash burns (superficial burns to exposed skin caused by the radiant and convective heat of the explosion. Methaemoglobinaemia due to poisoning by dinitrobenzene or potassium perchlorate (components of WWI munitions) [4] Acute septecemia melioidosis due to inhalation of soil particles contaminated with Pseudomonas pseudomallei [12] Explosions are associated with a high incidence of psychological sequelae in injured and uninjured survivors.

In the textbook *Disaster Nursing*, Sacco [8] adds a fifth category of blast injury, Quinary, which "Result from additives to explosives devices and hyperinflammatory states post-explosion," adding that "Further investigation into this type is required." Sacco goes on to include radiation exposure and its subsequent damaging effects as a possible additive to an explosive device.

Horrocks notes that the least-understood of these categories are the primary injuries, those that are due to the shock wave passing through the body. This shock wave can collapse hollow organs (e.g. the lungs) and it can shear connective tissue. The strength of the blast wave, while generally dropping in intensity with the inverse cube of distance from the source of the blast, can reflect from structures and surfaces, sometimes reinforcing each other to form a stronger shock than expected.

As the shock wave passes through the body it can cause localized damage that is frequently undetectable from the outside, including small hemorrhages, shear injuries, and so forth. Having said that, Horrocks also notes that the proximity to an explosion that would lead to blast injuries will likely also cause visible damage (lacerations, broken bones, traumatic amputation, and so forth). From the standpoint of medical responders, what is important to remember is that, for those who were closest to the blast, there might be severe internal injuries that cannot easily be detected.

It is worth noting that, while body armor might help to protect against secondary injuries, it provides limited or no protection against primary and tertiary injury. Thus, a police officer or military service person who is wearing body armor and who has few, if any, penetrating injuries might still have internal injuries from the blast.

In addition to injuries caused by the blast wave passing through the body, medical responders should be aware is the potential for crush injuries and compartment syndrome, which can be life-threatening if left unrecognized and untreated. Persons who have had heavy objects fall on them or who are pinned in place are most likely to suffer from crush injuries; compartment syndrome can arise from crush injuries as well as from broken bones, or even a partial loss of blood flow for an extended period of time (Fig. 6.1).

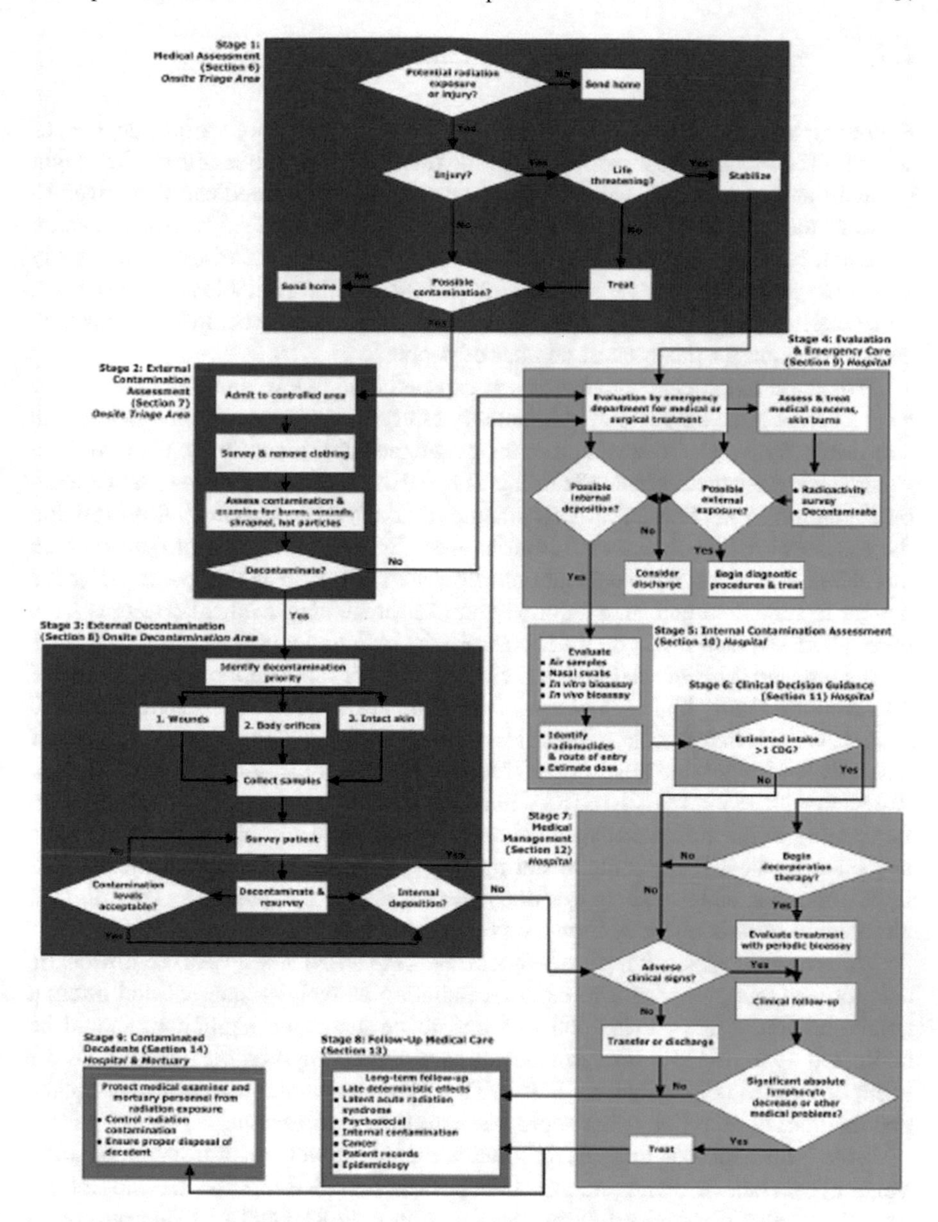

Fig. 6.1 Decision tree for management of contaminated persons, used with the permission of the National Council on Radiation Protection and Measurements [6]

6.3.4 Triage and Prioritizing Patients

Following a major disaster there will be people with injuries of various degrees of severity. Those who are seriously injured and for whom even the smallest delay might be significant *must* be treated first; those who are lightly injured can likely wait to be seen; those who are likely to die even with the best medical treatment cannot be saved, but can be made more comfortable. The purpose of triage is to quickly determine which category each person falls into so that the available resources (time, personnel, treatment rooms, medications, etc.) can be used to accomplish the greatest amount of good for the greatest number of people.

In the case of patients who have been exposed to high levels of radiation or who have taken in large amounts of radioactivity rarely succumb to radiation sickness in less than a few weeks to several months post-exposure. Accordingly, the Radiation Emergency Assistance Center/Training Site (REAC/TS) and the National Council on Radiation Protection and Measurements (NCRP) both recommend prioritizing the treatment of life-threatening injuries first. In fact, the very first questions on the NCRP Radiation Patient Treatment flowchart [6] address this point. After the patient is stabilized and their injuries and life-threatening medical concerns have been addressed then it will be time to assess the radiological concerns.

Having said this, internal radioactivity can call for rapid action if the amount of intake is sufficiently high—in excess of 10 Clinical Decision Guidance levels, or 10 CDG (1 CDG is that amount of internal radioactivity that will produce an exposure of 250 mSv—25 rem—the lowest dose that can produce clinically significant effects). A person with an intake of 10 CDG will receive a radiation exposure that can be life-threatening in the absence of medical care, especially in combination with injuries and other medical issues. Administering decorporation agents such as those noted in the following table can help to reduce internal radiation exposure by speeding the elimination of radionuclides from the body (Table 6.1).

With regards to prioritizing radiological risks, the NCRP and REAC/TS flowcharts call for evaluating exposure to external radiation as well as external and internal contamination. Any of these that is found to be medically significant should be addressed appropriately through medical care and appropriate medications. In the event of internal contamination, REAC/TS recommends minimizing further uptake and facilitating excretion using appropriate decorporation agents.

More information on these decorporation agents, as well as information on additional agents can be found, among other places, in NCRP reports 161 and 166 [6, 7], and can also be provided during consultation with REAC/TS in the event of an uptake. Medical response to radiological and nuclear attacks will be discussed in greater detail in Chap. 17.

Table 6.1 Recommended decorporation agents for several elements [6, 7]

Element	Agent(s)	Comments
Iodine	Potassium Iodide	Not needed if older than 40 Not needed if averted dose is less than 50 mSv Lower doses for children and babies
Cobalt	Calcium EDTA Calcium or zinc DTPA	Off-label use for all of these, there is no FDA-approved treatment
Cesium	Potassium ferrocyanide	Administer orally
Radium/Strontium	Aluminum hydroxide Barium Sulfate Sodium alginate	Monitor blood counts as Ra and Sr are bone-seeking elements
Uranium/Actinides	Sodium bicarbonate Calcium or zinc DTPA	Sodium bicarbonate makes urine alkaline to mobilize uranium Administer via IV or inhalation
Contaminated wounds	Calcium or zinc DTPA	Consider flushing contaminated wounds with DTPA to help remove contaminants Remove particles with tweezers if possible

References

1. Chaffee M (2009) Willingness of health care personnel to work in a disaster: an integrative review of the literature. Disaster Med Public Health Preparedness 20:1(42–56)
2. Horrocks C (2001) Blast injuries: biophysics, pathophysiology and management principles. J R Army Med Corps 147(1):28–40
3. Korir G, Karam P (2018) A novel method for quick assessment of internal and external radiation exposure in the aftermath of a large radiological incident. Health Phys 115(2):235–261
4. Laure P (1993) Methaemoglobinaemia: an unusual case report. Intensive Care Med 19:124
5. Martens K, Hantsch C, Stake C (2003) Emergency preparedness survey: personnel availability and support needs. Ann Emerg Med 42(Suppl 1):389
6. National Council on Radiation Protection and Measurements (2008) Report No 161: management of persons contaminated with radioactivity. NCRP, Bethesda MD
7. National Council on Radiation Protection and Measurements (2010) Report No 166: population monitoring and radionuclide decorporation following a radiological or nuclear incident. NCRP, Bethesda MD
8. Sacco T (2019) Traumatic injury due to explosives and blast effects. In Veenema (ed) Disaster nursing and emergency preparedness for chemical, biological, and radiological terrorism and other hazards. Springer Publishing Company, New York
9. Thompson D, Church C (2001) Prussian Blue for treatment of radiocesium poisoning. Pharmacother J Hum Pharmacol Drug Ther 21:1364–1367
10. US Environmental Protection Agency (2019) Federal guidance Report No. 15: external exposure to radionuclides in air, water and soil. US EPA, Washington DC
11. US Health and Human Services. Radiation Emergency Medical Management (REMM) website. https://www.remm.nlm.gov/. Accessed 11 Nov 2020
12. Wang C, Yap B, Delilkan A (1993) Melioidosis pneumonia and blast injury. Chest 103:1897–1899

Chapter 7
Physical Effects of Radiological Weapons

On March 11, 2001 a portion of the floor of the Pacific Ocean was ripped by one of the largest earthquakes ever recorded. The ensuing tsunami hammered the Japanese coastline, obliterating neighborhoods and industrial districts, and killing over 15,000 people who drowned, were buried by sediments, or were swept out to sea. Among other things, this tsunami also set in motion a chain of events that led to a series of nuclear reactor accidents at the Fukushima Dai'ichi reactor site; as a result, three nuclear reactors melted down, releasing radionuclides into the atmosphere and into the ocean. The total death toll due to radiation exposure during the emergency and for the five decades following the accident was estimated by the World Health Organization to be zero; to date there has been one death attributed to radiation exposure by the Japanese government, although the WHO noted that "the time for onset was shorter than the minimum latency period for radiation-induced leukemia" [1].

In spite of this great disparity between the toll taken by the tsunami compared that from radiation, the overwhelming perception on the part of the general public is that of a radiological disaster with little to no consideration of the physical damage and the fatalities caused by the earthquake and tsunami. In the aftermath of a radiological attack, while the radiological aspects of the attack will likely dominate the public discussion and perception in a similar manner, we cannot lose sight of the fact that an attack has taken place—very possibly involving large quantities of explosives—and that the physical effects of this attack must be considered during the initial response as well as during subsequent recovery efforts. At the very least, it would be a shame for an emergency responder to be so intent on monitoring radiation dose rates that they walk in front of a truck or step on a live electrical cable.

P. A. Karam, *Radiological and Nuclear Terrorism*,
Advanced Sciences and Technologies for Security Applications,
https://doi.org/10.1007/978-3-030-69162-2_7

7.1 Blast

We are, sadly, all too familiar with the effects of blast, be it from an improvised explosive device, a belt or vest worn by a suicide bomber, a letter or package sent by mail, or a vehicle filled with explosives. While explosions can form in solid, liquid, or gas the most common form is a solid.

Explosives consist of a fuel mixed with an oxidizer. When the explosion is initiated the fuel and oxidizer react, producing gas and releasing energy. One kilogram of TNT, for example, will produce 825 L of explosion gases and produces 4.184 MJ of energy when detonated [2]; when this gas is heated by the energy released by the chemical reaction the volume expands even further. This is the origin of the blast from an explosion [2–4].

If the chemical reaction passes through the explosive relatively slowly (lower than the speed of sound) the material is called a low explosive; high explosive materials are those in which the chemical reaction proceeds through the material at supersonic speeds. Explosives in which the chemical reaction moves more rapidly and in which more energy is released per chemical reaction tend to be more powerful; this can be increased further by the addition of aluminum powder or other enhancing materials.

As the chemical reaction progresses the expanding hot gas creates a pressure wave that expands into space. The sudden increase in pressure and temperature creates a shock wave—a large and abrupt change in pressure in a very small distance—that causes damage as it encounters objects in its path. If those objects are relatively strong compared to the energy in the shock wave then they will be physically displaced; if the objects are relatively weak compared to the strength of the shock wave then they will be torn apart. The strength of the shock wave drops roughly as the inverse cube of distance from the source of the explosion.

7.2 Damage to Structures

When the shock wave from an explosion impacts a structure, it will impart pressure across the entire cross-sectional area of the structure. Due to the relatively large cross-sectional area of even a small structure, this can exert a large amount of force; a pressure of only 10 kPa (1.45 psi) applied to a wall three meters high and 10 m in length (30 square meters) will exert a force of 300,000 N—more than 30 tons of force—imparted over a fraction of a second. This is enough force to shatter glass and to collapse many ordinary building materials.

As noted above, an explosion of any significant size can place a tremendous lateral load on a structure. In addition to the damage caused to the walls and windows (to be discussed below) this load can also weaken structural members and can cause a building to collapse entirely or partially; it can also permit a building to remain standing while being unstable and prone to collapse at a later time, during emergency response activities. Those responding to any bombing attack, regardless of the presence (or absence) of radioactivity must be aware of the potential for structural damage that can make the building unsafe to enter and unsafe to work in the vicinity of [5].

Glass that is blown into a room can pose a hazard to those inside the room—it can cause lacerations ranging from minor to life-threatening. But that is not the only hazard posed by glass; the blast can also cause sheets of glass on the exterior of newer buildings to fall to the ground where it can injure or kill people at street level—not just by causing lacerations, but also due to the weight of heavy panes of glass falling from a height. A pane of double-glazed residential glass weighs about 20 kg per square meter [16] so a window that is two meters wide and three meters in height will weigh 120 kg. The impact of even a portion of this pane falling from a height of several tens or hundreds of meters can be fatal.

Similarly, decorative and other features can be blown off by the blast, or can be weakened or loosened enough to break loose and fall at a later time. These, too, can pose a threat to anyone below.

Thus, blast damage to structures can pose a risk to persons inside the buildings who might be injured by flying glass or who might be trapped in a structure that collapses at some time after the explosion. In addition, there is a risk to those outside the building who must also be concerned about unstable structures and broken glass (in this case, falling on them from above), as well as pieces of the building falling from above (Fig. 7.1).

7.3 Damage to Utilities

The blast and shock from a bomb affect more than buildings; they can topple utility poles or can be transmitted through the ground to damage or rupture subsurface utility lines. In addition, collapsing buildings can bring down utility poles or, if they are sufficiently heavy and fall far enough, debris can penetrate far enough into the ground to sever utilities.

Explosions do more than create shock and blast waves, they also excavate materials they are in contact with—they create craters. The volume of material excavated by an explosion is roughly proportional to the energy released by the blast [7] so the craters formed by large explosions are deeper and wider than those formed by smaller explosions in the same material. The other controlling factor is the strength of the material being cratered; stronger, denser, more consolidated, and more competent materials require more energy to produce a crater of a given size compared to materials that are lighter, weaker, or more granular. Craters are more than just holes in the ground; a crater that runs beneath a building can contribute towards destabilizing the building, while the formation of the crater itself can also damage nearby utility lines, or lines that pass through the volume of the crater. Thus, a crater can produce a risk (or collection of risks) that will continue until they are addressed and resolved.

A utility line that passes through the volume of a crater might or might not be ruptured by the explosion that formed the crater. However, even if the line appears to

Fig. 7.1 **a** The Alfred P. Murrah Federal Building following the Oklahoma City attack. The labels refer to specific support columns [6]. **b** Diagram of the crater left by the Oklahoma City truck bomb [6]

be intact it might well be damaged or weakened by the explosion. Not only that, but a pipe (for example) that had been supported along its length by the soil or trench in which it had rested will now be unsupported across the diameter of the crater; this places additional stress on the line that can cause it to fail, causing additional dangers.

- Electrical lines can create electrocution hazards at the site of the break as well as in any location that is electrically connected to the line (e.g. a pool of water or length of metal rebar). In addition, damage to electrical lines can cause longer-term public health problems and can impair response efforts until generators can be brought to the scene.
- Steam line ruptures can cause serious burns from leaking steam. In addition, many steam lines were once insulated with asbestos-bearing lagging and steam line ruptures have introduced asbestos into the atmosphere and surrounding areas.
- Water and sewer line ruptures can cause flooding, can undermine pavement leading to later collapse, and can raise public health concerns in the days, weeks, or months following the explosion. Damage to water lines might also impair fire-fighting and decontamination or contamination control efforts.
- Natural gas line ruptures can ignite or explode (especially if they are in proximity to downed or ruptured electrical lines), posing a serious threat to the health of anyone in their vicinity. This can also deprive those living in the area of heat, hot water, and the ability to cook until the lines can be repaired.
- Subway lines, pedestrian tunnels, utility access tunnels, and the like can be breached, weakened, or collapsed. This can put people at risk directly and can also make it possible for groundwater to flood the tunnels over a longer period of time. In addition, because subways are generally electrically powered, and any tunnels meant to be accessed by members of the public will be lighted, there will be electrical lines running through the tunnels that can be damaged or short-circuited by water flooding the tunnel, causing additional risks to people who might be in the tunnels or who might enter in order to conduct emergency response activities.

Thus, not only do damaged or ruptured utility lines pose a risk at the site of the attack, but the loss of these utilities can also complicate emergency response efforts in the short term and can lead to longer-term public health concerns for those affected.

7.4 Fires

As noted above, natural gas lines can rupture and ignite, causing fires, but the explosion itself can also ignite fires. Having said this, high explosives (those in which the detonation velocity is greater than the speed of sound) "are primarily intended to cause damage through blast and…fragmentation" [8] and are less likely to ignite fires compared to low explosives such as black powder or ANFO (ammonium nitrate mixed with fuel oil) or explosives that consist of fuel sprayed into the air and ignited.

Fires pose risks beyond the obvious concerns about burns, smoke inhalation, destruction of property, and so forth. Fires can also create winds that can resuspend radioactive particles, the rising thermal columns of air from fires can inject these particles into the atmosphere, and winds can spread this contamination far downwind. Some studies (e.g. [9]) concluded that appreciable amounts of contamination can be spread a few tens of kilometers downwind if the contamination is lofted high enough into the atmosphere to be caught by upper-level winds; this can be accomplished by the blast itself or by fires started by a suitable explosion.

The water used to combat fires can also help to spread contamination; radioactivity can be washed down streets and into the storm sewers by the water used for firefighting, contaminating the sewers and (depending on the sewer system design) possibly also contaminating waste water treatment plants. Fortunately, studies have shown that properly working water treatment plants are very effective at removing radioactivity from waste water, but the plants themselves will become contaminated. Care must also be taken to segregate the solids produced by waste water treatment to prevent contamination of the landfills or fields to which these solids are removed for disposal [10].

7.5 Flying Objects

The blast from a bomb is capable of throwing objects long distances at high speeds; these objects can be pieces from the bomb itself and whatever it is that contains the bomb (e.g. suitcase, car, trash can, etc.), nearby objects that are picked up or blown off of nearby objects by the blast (e.g. street signs, soda cans, pieces of asphalt and brick, etc.), and any other items that can be mobilized by the blast wave. There have even been examples of teeth and fragments of bone from a suicide bomber becoming shrapnel [11]. The distance these objects can be thrown depends on the strength of the explosion, the weight of the object, and their aerodynamic properties; a small piece of dense metal will be propelled further than, say, a street sign of the same weight but with a larger cross-sectional area. For this reason, a number of organizations have developed recommendations for safe stand-off distances from a variety of bomb sizes. These distances range from about 20 m (the mandatory evacuation distance from a 2-kg pipe bomb) to about 3 km (the preferred evacuation distance from a 30-ton truck bomb [17, 18] (Fig. 7.2).

Among the materials that can be scattered by an RDD are fragments of the source itself as well as radioactive powders or liquids (depending on the physical form of the source(s) used. Thus, in the aftermath of an attack using an explosive RDD, personnel approaching the scene should enter from the upwind direction or laterally if possible and should monitor for elevated radiation dose rates when they are at least as far away as the distances noted on this table. The force of the explosion can also inject these powders or liquids into the air where they can be spread downwind; this will be discussed in greater detail in the following section.

BOMB THREAT STAND-OFF CARD

Threat Description	Explosives Capacity	Mandatory Evacuation Distance	Shelter-in-Place Zone	Preferred Evacuation Distance
Pipe Bomb	5 lbs	70 ft	71-1199 ft	+1200 ft
Suicide Bomber	20 lbs	110 ft	111-1699 ft	+1700 ft
Briefcase/Suitcase	50 lbs	150 ft	151-1849 ft	+1850 ft
Car	500 lbs	320 ft	321-1899 ft	+1900 ft
SUV/Van	1,000 lbs	400 ft	401-2399 ft	+2400 ft
Small Delivery Truck	4,000 lbs	640 ft	641-3799 ft	+3800 ft
Container/Water Truck	10,000 lbs	860 ft	861-5099 ft	+5100 ft
Semi-Trailer	60,000 lbs	1570 ft	1571-9299 ft	+9300 ft

Fig. 7.2 Bomb threat stand-off card circulated by the Department of Homeland Security [17]

Medical responders should be aware that patients might contain small pieces of radioactive source within their bodies that can emit high radiation dose rates. Cobalt-60 source pellets, for example, are only a few mm in diameter but can contain several tens of MBq of activity, producing radiation dose rates in excess of 10 mSv/hr at a distance of one meter. This does not pose a threat to medical caregivers, although it can cause radiation burns and tissue destruction in the immediate vicinity of the embedded pellets.

The Oklahoma City Bombing

On April 19, 1995 Timothy McVeigh detonated a bomb that consisted of approximately 2200 kg of ANFO just a few tens of meters from the Alfred P. Murrah Federal Building in Oklahoma City.

The force of the explosion blew off the building's façade and forced the lower floors to flex upwards, weakening them and contributing to their collapse. About one third of the building was destroyed and 168 people died from the explosion and from the collapse of the building.

The bomb produced a crater that was nearly 10 m in diameter and about 2.4 m in depth, and it damaged 324 buildings, breaking glass in 258 of them. There were 851 dead and injured, 447 of whom were treated at local hospitals. Injuries included lacerations and other soft tissue injuries (403), fractures and dislocations (78), head injuries (100), severe lacerations (20), eye injuries (70), and burns (9). Broken glass accounted for 38% of the 2115 injuries for which the cause could be determined, more than any other cause [12].

Parts of the truck in which the bomb was hidden were found up to several hundred meters away.

7.6 Contaminated Area and Infrastructure

One of the major physical effects of a radiological attack will be the areas that will be off-limits due to radiation dose rates and contamination levels. This can include residential areas, bridges, roads, tunnels, stores, government buildings, transportation hubs and equipment, port facilities, and much more. The loss of these facilities, whether due to a radiological attack, extreme weather, or any other event will be disruptive for some or all of a city's residents.

Let's consider an explosive RDD – a "dirty bomb"—that is set off in a crowded area in a city center. We will assume that the radioactivity itself is in a dispersible form—a powder or a liquid—to facilitate its spread. The majority of the radioactivity will remain in that form, but some will work its way into cracks or crevices within the container and some might clump together. When the explosives are detonated the blast wave will pass through the container in which the radioactive material is held; the heat and pressure will blast the contamination into the atmosphere along with pieces of whatever containers hold the explosives and the radioactivity. As this is happening, most of the radioactivity will be dispersed into the atmosphere, but some will fuse with the materials blown apart by the explosion—the glass, plastic, metal, and other materials of which the containers, the weapon, and (if appropriate) the vehicles are constructed. This means that there will be an assortment of particle sizes, a wide array of fragments of varying size and composition, and they will be distributed across a variety of distances from the scene of the explosion. Larger particles will tend to follow a ballistic trajectory and will typically travel up to a few hundred meters unless they strike a building, vehicle, vegetation, or the like, with the larger particles (several microns and larger in size) tending to settle to the ground within several minutes [13].

Smaller particles are light enough to remain airborne for longer periods of time and will tend to travel with the air currents. While this will tend to be downwind, in an urban environment the winds can swirl around buildings and eddies can carry particles upwind when the wind is blowing at an angle to the city's grid of streets and buildings. Aerosol-sized particles can travel a few to several tens of kilometers downwind, spreading contamination wherever the plume settles to the ground.

Studies performed at Sandia National Laboratory [9] indicate that, for an explosive RDD detonated at Wall Street in New York City, areas requiring remediation to meet regulatory radiation exposure limits would cover the southern tip of Manhattan and measurable contamination could extend as far as JFK airport to the east, a distance of more than 30 km, contaminating not only buildings, but public spaces, municipal buildings, grocery stores, apartment buildings, subway stations and trains, roads, bridges and tunnels, and more. The Sandia study estimated that nearly 200,000 people would be required to relocate due to regulatory requirements and several hundred thousand additional people would be living and/or working in areas that were measurably contaminated, albeit to levels that did not call for remediation; Fig. 7.3 gives a general idea of the extent of the plume from such an attack. The societal impact of this will be discussed in greater detail in Chap.8.

An explosive RDD is obvious and we can expect that the radiological nature of the attack will be discovered within several minutes of the explosion in most major cities. But it is also possible to distribute radioactivity without setting off a bomb—this is what Peter Zimmerman called a "smokey bomb;" a covert release of radioactivity. In the event of such an attack the city might not realize that an attack has even taken place for days or even weeks after the release has taken place. In this time, the contamination can spread not only throughout the city that was attacked, but can spread to other cities around the world as well. In the aftermath of the 2006 assassination of Alexander Litvenenko in London not only were over 30 locations in London found to be contaminated, but several commercial aircraft also

Fig. 7.3 An example of contaminated areas and the plume from an RDD set off on Wall Street

had measurable levels of contamination onboard and passengers brought contamination to over 50 international destinations as well [14, 15]. It should be noted that this wide spread of contamination was the result of the incidental spread of a small amount of radioactivity that spread from two men who poisoned a third; a deliberate covert attack against a major city waged with malicious intent would almost certainly be far more widespread.

The effect on transportation infrastructure can be especially disruptive, especially in cities that rely heavily on mass transportation, such as London and New York City. In many cities, the train stations are connected to the outside atmosphere by grates and the design of the train stations and the piston action of trains traveling along the tracks pumps air through the system, providing fresh air for passengers and workers. If this air is contaminated, this same piston effect will be quite effective at pushing radioactivity throughout the subway system and spreading it to distant parts of the city. This can not only spread contamination in directions that might seem counter-intuitive, but can also cause the subway system to be shut down, severely impacting the ability of citizens to travel from home to work and back, to visit friends, to attend events in other parts of the city, and so forth.

7.7 Radioactive Debris

After the emergency phase is over the city will be faced with cleaning up the radioactive contamination. The details of this remediation will be discussed in Chaps. 20 and 21; here we will discuss factors related to the scale of the response that will be required. The primary factors that will impact future remediation efforts include:

- The form and physical volume of radioactive materials used in the device,
- The degree of dispersibility,
- The strength of the explosion (for an explosive RDD),
- The area in which the device is used, and
- The extent of the contamination.

Radioactive sources come in a variety of physical and chemical forms. A fine powder or liquid will disperse far more readily and more evenly than will a radioactive metal alloy; powder that is sealed within a welded capsule will not disperse as readily as will the same powder that is in an unsealed form. Soluble materials are easily spread by rain or by water used to fight fires, contaminating the sewer system, waste water treatment plants, parks, rivers, and will lead to a longer and more expensive remediation in the future.

All else being equal, a greater volume of radioactive material will contaminate a larger area than will the same source activity that is physically smaller. For example, one gram of pure ^{60}Co will have a volume of about 17 μl. If the average particle diameter is 5 microns then each particle will have a volume of about 5×10^{-11} μl. Thus, a source that is pure ^{60}Co will contain about 350 million particles. By comparison, if the ^{60}Co is well-mixed and alloyed with 1 kg of other metals and is divided into particles of the same size there will be 350 billion particles and a much larger area will be contaminated by the same RDD, albeit to a lower concentration of radioactivity per unit area.

The area in which an RDD is used will also affect its impact. The release, covert or explosive, of radioactivity in an area surrounded by skyscrapers will result in the contamination of many expensive buildings and will produce a larger economic impact than a release in a less-built area. Winds can also be redirected and concentrated when passing between skyscrapers—the "urban canyon" effect—so radioactivity released in an area filled with skyscrapers will end up being transported along the axis of the urban street grid (instead of in the direction of the prevailing winds) and will travel further at higher velocities than if released in a more open area. At the same time, the presence of tall buildings can absorb the majority of the blast wave and contamination, reducing the extent of blast damage and the spread of large concentrations of contamination downwind.

Another factor to consider is the effect of contamination on the buildings themselves. In particular, contamination on building exteriors is relatively easy to wash off and might even be ignored at high elevations. However, contamination that enters the building ventilation system will spread throughout the structure, contaminating the ventilation system and possibly the entire interior volume as well if ventilation is not turned off quickly. Given the high cost of real estate and of construction, this can lead to a very high economic impact if valuable real estate must be removed from service, demolished, disposed of as radioactive waste, and rebuilt.

The extent of radioactive contamination—the buildings and land contaminated by radioactive debris—will, to a large part, determine the overall financial and societal impact of an RDD. If the contamination does not spread far, if few structures are contaminated and if the contamination does not contaminate a large amount of a city's infrastructure then the overall impact of the RDD will be far lower than if hundreds of buildings must be decontaminated and demolished at a cost of hundreds of billions of dollars.

References

1. World Health Organization (2013) Health risk assessment from the nuclear accident after the 2011 Great East Japan Earthquake and Tsunami, based on a preliminary dose estimation. World Health Organization, Geneva Switzerland
2. Meyer R, Kohler J, Hormburg A (2007) Explosives, 6th edn. Wiley, New York
3. Akhavan J (2011) The chemistry of explosives, 3rd edn. Royal Society of Chemistry, Cambridge UK
4. Davis T (1943) Chemistry of powder and explosives. Wiley, New York
5. Federal Emergency Management Agency (2003) Primer for design of commercial buildings to mitigate terrorist attacks. FEMA, Washington DC
6. Federal Emergency Management Agency (1996) The Oklahoma City bombing: improving building performance through Multi-Hazard Mitigation. FEMA, Washington DC

7. Vortman L (1977) Craters from surface explosions and energy dependence—a retrospective view. In: Roddy D, Pepin R, Merrill R (eds) Impact and explosion cratering. Pergamon Press, New York
8. Geneva International Centre for Humanitarian Demining (2017) Explosive weapons effects. GICHD, Geneva
9. Trost L, Vargas V (2020) Economic impacts of a radiological dispersal device. Presented to the National Academies of Science, Albuquerque NM
10. Washington State Department of Health (1997) The presence of radionuclides in Sewage Sludge and their effect on human health. Division of Radiation Protection, Olympia WA
11. Patel H, Dryden A, Gupta A, Stewart N (2012) Human body projectiles implantation in victims of suicide bombings and implications for health and emergency care providers: the 7/7 experience. Ann R Coll Surg Engl 94(5):313–317
12. Shariat S, Mallonee S, Stephens Stidham S (1998) Summary of reportable injuries in Oklahoma: Oklahoma City bombing injuries. Oklahoma State Department of Health, Oklahoma City
13. Harper F, Musolino S (2007) Realistic radiological dispersal devices hazard boundaries and ramifications for early consequence management decisions. Health Phys 93(1):1–16
14. Harrison J, Fell T, Leggett R, Lloyd D, Puncher M, Youngman M (2017a) The polonium-210 poisoning of Mr. Alexander Litvinenko. J Radiat Prot Res 37:1(266–280)
15. Harrison J, Smith T, Fell T, Ham G, Haylock R, Hodgson A, Etherington G (2017b) Collateral contamination concomitant to the polonium-210 poisoning of Mr Alexander Litvinenko. J Radiat Prot Res 37(4):837–851
16. Glass Technology Services (2020) Glass weight calculator. https://www.glass-ts.com/glass-weight-calculator. Accessed 12 Nov 2020
17. National Ground Intelligence Center (2020). Improvised explosive device (IED) Safe standoff distance cheat sheet. Department of the Army, Washington DC. Available online at https://www.hsdl.org/?view&did=440775, Accessed 12 Nov 2020
18. US Department of Homeland Security, Office of Bombing Prevention. Bomb threat stand-off chart

Chapter 8
Societal Impact of a Radiological Attack

Any radiological attack is likely to have a profound impact on society; the mere mention of radiation will frighten the population (and likely politicians as well), contaminated areas will likely be off-limits until they can be decontaminated and access restored, and there will be a tremendous economic impact at all levels of society from individuals through the local and national governments.

Governments will also need to consider both short- and long-term health effects on their populations; how many were injured by the blast, how many will suffer radiation sickness, how many might develop cancer over the decades to come, and who will pay for all of the testing and medical care over as long as a half-century after the event.

8.1 Short-Term Impact

For the purposes of this discussion, "short-term" will refer to events that transpire during the period of time from the recognition that an attack has taken place through the days or the first few weeks that pass until the full extent of contamination is known and those areas delineated with rescue operations completed. This will be the period with the most confusion and upheaval, the period during which the responders will be most active, the public will be the most frightened, and government is likely to be the most confused and disorganized. As the full extent of the event becomes known and as the public, responders, and government begin to adjust to the attack and to plan their respective actions and work to recover from the attack—as individuals, businesses, and governments—will begin.

P. A. Karam, *Radiological and Nuclear Terrorism*,
Advanced Sciences and Technologies for Security Applications,
https://doi.org/10.1007/978-3-030-69162-2_8

8.1.1 Public Health

During and after the emergency phase of a radiological attack the primary public health concerns will revolve around providing care for the critically injured as discussed in Chaps. 6 and 17. This is likely to be challenging as hospitals close to the scene of the attack are likely to be overwhelmed with the first victims to be evacuated from the scene as well as by persons who self-evacuate and transport themselves to the hospital. In the aftermath of the attacks against the World Trade Center on September 11, 2001 public health, emergency response, and medical caregivers reported to the author that not only did large numbers of victims arrive at nearby hospitals by foot, automobile, or mass transit, but many who were relatively lightly injured actually arrived at medical facilities more quickly than did the critically injured because they did not require rescue, evaluation, and transport.

In New York City, for example, the primary radiological public health objectives are to care for those in the immediate vicinity of the dispersal, whether explosive or covert. Taking steps to reduce inhalation and ingestion, moving members of the public out of the path of airborne radioactivity, and telling those downwind to shelter (go indoors, close doors and windows, and wait until the plume has settled to the ground), and taking steps to determine the plume "footprint" will help to minimize radiation exposure to those who are downwind but not directly affected by the explosion.

Once these immediate needs are addressed—caring for those who require medical attention and minimizing radiation exposure to those more distant from the release—public health personnel can begin to pivot towards considering long-term health concerns.

8.1.2 Economic

A radiological attack is likely to have a massive economic impact, but this will play out over the course of weeks to decades after the attack; in the short term, during the emergency phase, the economic impact is likely to be far lower. In such a short time frame the major economic impact for the government is likely to be overtime paid to emergency responders and obtaining the supplies needed to stabilize the scene and to care for the injured as well as those who need shelter because their homes were destroyed or are inaccessible.

In the days following an attack those whose homes were destroyed or made inaccessible will need to pay for lodging and for items they will need on a day-to-day basis; in the absence of assistance from the government this can have a negative effect on their savings and net wealth. Businesses in the area will suffer as well when their customers (and possibly employees) are unable to visit. In fact, the issue of inaccessibility is likely to continue for months to decades, depending on a number of factors, which are discussed below. In the short term these businesses will, like the individuals affected, have to spend their savings (if any) to continue paying necessary expenses until they can reopen or receive assistance from the government. However, the short-term economic effects should be fairly limited, if only due to the short period of time that this phase is expected to last.

Estimating evacuation-related fatalities

According to information collected by the US Department of Transportation [1] Americans face a fatality rate of about 1.13 deaths for every 100 million miles (160 million km) driven. Keeping this in mind, consider the impact of evacuating New York City in the aftermath of a nuclear reactor accident or a terrorist attack against New York City.

If we assume the entire population of New York City (about 8.5 million people) evacuate a distance of 100 miles with an average of three people per vehicle we can expect 2–3 fatalities to occur while evacuating and the same number when returning home.

This quick calculation is based on normal drivers, normal weather, and driving under normal circumstances. If the weather is bad, the drivers are angry, rushed, or panicked, and/or the roads are jammed with traffic the actual number of fatalities is likely to be higher.

8.1.3 Psycho-Social

In spite of exciting optimism and enthusiasm when first discovered, radiation and radioactivity were known to cause deterministic health effects within a few years of their discovery and the tragic fate of the "radium girls" (young women who ingested radium while "pointing" brushes to paint the dials of self-luminous watches and instruments) brought many of these health effects to the attention of the general public in the 1920s. With the use and later testing of nuclear weapons in the 1940s, 1950s, and 1960s the public began to associate radiation with nuclear testing; this, combined with increasing environmental awareness and an almost total lack of knowledge about radiation on the part of the public helped to fuel increasing fears that morphed, in some cases, into outright phobia [2]. Fear of radiation can cause the public to respond inappropriately to radiological and nuclear events. After the Three Mile Island accident, for example, approximately 144,000 residents voluntarily evacuated the area in spite of the fact that the evacuation order only recommended that pregnant and nursing women leave the area [3]. While this did not lead to any noticeable increase in traffic fatalities, this was not the case in aftermath of the Fukushima accident in which, of over 130,000 evacuees, over 1600 died [4].

The short-term psycho-social effects do not stop when evacuees reach their destination; they will need to live somewhere until they can return home and they can face social stigmatization owing to the radiophobia of their new neighbors. The author was part of a three-person team that travelled to Japan about one month after the Fukushima accident; while there, members of the team met with people at shelters and with evacuees. Many of those in the shelters were depressed and/or anxious, while some of the evacuees mentioned that those living near them were voicing complaints about noise, cooking odors, and so forth that appeared to be proxies for an underlying fear of the radiation to which the evacuees had been exposed. A similar social stigmatization (as well as a number of other psychosocial impacts) has also been noted to affect survivors of radiation and nuclear reactor accidents around the world [5–9].

To help address these, and related, issues New York City emergency planning documents call for including mental health information and referrals among the services they plan to offer to those affected by a nuclear or radiological attack [10].

8.1.4 Access Restrictions to Impacted Areas

As with any other emergency it will be necessary to quickly delineate the affected area(s). One reason for this is for public health and safety, so that members of the public are evacuated from, and denied entry to areas that might pose a risk due to radiation, contamination, and other hazards (e.g. fire, unstable structures, downed or ruptured utilities, etc.); it also serves to keep members of the public from interfering with the emergency response efforts as well as preserving the area as a crime scene for future investigations.

The immediate aftermath of a radiological attack will include delineating the Hot Zone based first on default boundaries (e.g. 250 m in all directions and 2000 m downwind, as suggested by New York City's Radiological Response and Recovery Plan [25]) and, when more information becomes available, adjusted according to procedures developed by the US Department of Homeland Security [11]. However, there are also non-radiological factors to consider that might call for adjusting boundaries further still. These factors can include the risk from secondary explosive devices, secondary attacks in the area of an explosion, risks from falling glass, damage to utility lines, the release of asbestos or other harmful materials into the environment, the presence of smoke or contaminant plumes, and so forth.

The tension in setting boundaries is that they should be expansive enough to encompass all potential sources of risk while remaining as compact as possible to facilitate securing them and to minimize the inaccessible area(s). It also makes sense to establish boundaries at "natural" locations (e.g. street corners versus the middle of the block) when possible.

Once boundaries have been established responders should also establish entry/exit points and, to the maximum extent possible, all traffic into and out of the controlled area should pass through these to ensure that all who enter the area have proper protective equipment and that all who exit are checked for contamination and decontaminated if necessary (except for the critically injured, who may be transported directly to the hospital for life-saving medical care).

Elected and appointed officials as well as the Incident Commander will also have to decide whether public health and safety is best served by initially sheltering in place or evacuating the public. While the initial impulse is normally to move people out of harm's way, it is important to remember that radiation is only one of multiple risks that would be faced during an evacuation; evacuation might not be a course of action that minimizes the overall risk to the public. For this reason, some cities have concluded that the immediate recommendation to members of the public should be to shelter indoors (provided the building is stable and not on fire) until conditions have stabilized to the point at which they can be evacuated safely.

8.2 Long-Term Impact

As serious as the initial impact of a radiological attack might be, such an attack is likely to continue to have repercussions for years or even decades, and the nature of these impacts is likely to change as time goes on. In the weeks, months, and years following an attack the focus of a city's response efforts will evolve away from stabilizing the scene and saving lives and towards attending to the long-term health and safety of residents as well as towards recovering from the attack and restoring access to the impacted areas.

8.2.1 Public Health

Many deterministic radiation injuries take days to weeks to manifest; thus, radiation sickness, skin burns, and the like can continue to appear for some time after an attack occurs, and even those who manifest deterministic effects early can continue to develop new symptoms for months or years [12].

In addition to the deterministic effects, leukemia and other blood cancers can begin to appear about five years or so after initial exposure with solid tumors beginning to arise about a decade post-exposure and continuing to appear decades later [13]. In this, the public health impact of a radiological attack is likely to parallel that of the attacks on September 11, 2001, which saw (and continues to see) a continuing stream of responders and members of the public being affected by the materials they inhaled on the day of the attack and the days following the attack.

Having said that, the rates of cancers in the aftermath of the Chernobyl, Three Mile Island, and Fukushima nuclear reactor accidents have not increased appreciably in the time since those respective accidents, and there was not a statistically significant increase in cancers among the lowest-dose cohorts in Hiroshima and Nagasaki [26] . This does not mean that there were no cancers induced by radiation exposure from those events, just that, if there were, they did not appear in high numbers. This is consistent with a statement in the 2008 report by the United Nations Scientific Committee on the Effects of Atomic Radiation in which it was noted that a study of the existing scientific and medical literature showed "…no evidence to support changing the conclusions (that) no effects are expected at chronic dose rates below 0.1 milligrays per hour or at acute doses below 1 gray to the most highly exposed individuals in the exposed population" [14].

8.2.2 Contamination

Contamination of medical equipment and facilities might also play a role in public health in the aftermath of a radiological attack. New York and other cities have realized the difficulty of attempting to confine the public to the immediate vicinity of the attack, releasing them only after checking them for contamination, and emergency responders are trained to take critically injured victims directly to the hospital without stopping to decontaminate them first. While this approach is likely to save the most lives, it also means that ambulances, emergency rooms, trauma bays, and operating rooms are likely to be contaminated and to require decontamination before they can be used to treat non-radiological patients. Decontamination resources (personnel, contractors, and equipment) are likely to be inadequate to address the city's decontamination needs; those hospitals with an in-house radiation safety program (typically large research hospitals) will likely be able to perform their own decontamination but smaller hospitals, urgent care centers, and clinics might be unable to see patients for a few to several weeks unless they receive permission to continue operating or unless they can install coverings on floors, furniture, and equipment to contain the contamination.

Yet another public health impact to consider regards the reproductive effects of a radiological attack. While the developing fetus is remarkably resistant to the effects of radiation exposure ([15], [24]) this is not widely known among members of the public nor among medical personnel; as a result it is not uncommon for pregnant women to be advised to have a therapeutic abortion following even relatively minor exposures to medical radiation [16]; this author has direct experience working with patients whose physicians have advised a therapeutic abortion following diagnostic x-rays that produced far too little radiation exposure to interfere with fetal development. In an article published by the Society of Nuclear Medicine about one year after the Chernobyl accident [17] the author states "According to the IAEA, an estimated 100,000–200,000 wanted pregnancies were aborted in Western Europe because physicians mistakenly advised patients that the radiation from Chernobyl posed a significant health risk to unborn children."

In the aftermath of a radiological attack, public health officials should expect an increase in the numbers of women seeking abortions and a decrease in births as a result of fear on the part of the parents and lack of information on the part of their physicians. This can be addressed in part by developing and distributing fact-based information resources for physicians and pregnant women both.

8.2.3 *Psycho-Social*

Radiological and nuclear events can traumatize the public even when radiation exposures are too low to cause harm. In fact, when the World Health Organization investigated the health of those living in the Chernobyl area in the first two decades after the accident [27] they found that the mental health impact of this event was greater than the medical effects. In particular, incidence of substance abuse, depression, anxiety, and the like were affecting not only those who were directly affected by the accident, but their children as well (possibly as a result of living with depressed, alcoholic, or suicidal parents).

As noted earlier, the psychological and social effects appeared in the aftermath of the 1987 radiological contamination accident in Goiania Brazil and following the nuclear reactor accidents in Fukushima Japan as well. This includes the social stigmatization noted above and the longer-term problems arising from the forced (albeit—hopefully—temporary) evacuation of large numbers of people, the loss of income for those whose workplaces are shut down due to being within the Hot Zone, fears of radiation and worries about long-term health effects, and more [18].

8.2.4 *Economic*

The long-term economic impact of a radiological attack can be tremendous. Zimmerman and Loeb [19] estimated that an RDD attack against a major city could cost tens to hundreds of billions of dollars; this would include the loss of tax revenue, the closure of businesses and loss of jobs, loss of access to critical infrastructure and residential properties, the decontamination of buildings and public spaces (streets, sidewalks, parks, etc.), remediation of environmental contamination, the disposal of radioactive waste from decontamination and remedial activities, and more.

The cost of any of these factors can vary substantially. For example, a building that is contaminated only on the exterior façade can be decontaminated using any of a number of established techniques, or the decision can be made to remove loose contamination but to let fixed contamination remain on upper floors if it does not add appreciably to radiation exposure to those working in the building or passing by at ground level.

On the other hand, many buildings have ventilation intakes at ground level and these can introduce radioactivity into the building's interior. There, the entire ventilation system and the entire interior can become contaminated; decontamination in this case could require removing the entire ventilation system, all of the furniture and carpeting, interior walls, and any other porous surfaces inside the building. This can increase the time that a building cannot be used from months to years—if decontamination is even possible. It can also increase the cost of decontamination by two to three orders of magnitude, given the greater length of time and the larger amount of radioactive waste that will be generated. In some cases, it might actually be more cost-effective to demolish and replace a building that is so thoroughly contaminated.

Among the most significant decisions that must be made is that of decontamination standards and this decision will be central in determining many aspects of the societal and economic cost of a radiological attack. As a rough rule of thumb, the cost of remediation increases as the square of the amount of risk reduction. For example, reducing the contamination levels (and, thus, risk) by a factor of two will cause costs to increase by a factor of four or greater; reducing contamination levels by a factor of three will increase costs by a factor of nine. This is due to the increased time and effort required for the additional cleanup as well as the additional volume of waste generated [20].

But we must also factor in the societal cost of having buildings, parks, roads, and infrastructure inaccessible or unavailable for use for a longer period of time. For example, consider the impact on New York City from a major subway station and associated train lines being closed for several years in order to permit decontamination to levels indistinguishable from background, versus taking only one year to decontaminate to levels that would expose workers to as much as 1 mSv annually. An analogous situation occurred in the aftermath of Hurricane Sandy in 2012 and some train lines were re-routed or closed down for only a year—the total cost to the Metropolitan Transit Authority exceeded $5 billion, including repairs to stations, train lines, tunnels, and lost revenues [21]. The cost of these disruptions extended beyond the money needed for repairs and lost fares, it also included the cost of passengers' time necessitated by the disruptions; as one example, the commuting time from the author's home in Brooklyn into Midtown Manhattan increased by 30–60 min depending on the day and time one was commuting.

Remediation will go more quickly if cleanup standards are based on risk rather than on our technological ability to detect radiation. Using information provided in Federal Guidance Report #12 [22] one can calculate that the radiation exposure from an area contaminated to a level of 10 dpm cm^{-2} (a common US limit to restore an area to unrestricted use) is less than 20 nSv hr^{-1}, which is only a small increase over normal background radiation levels. Nevertheless, this level of contamination is easily detected using handheld instruments and it is not unreasonable to think that members of the public might refuse to return to areas contaminated to these levels to work or to live. The fears of the public might well exacerbate the economic impact of such an attack.

Additional economic impacts can be the result of peoples' preferences and habits. Becker and Rubinstein [23] address one aspect of this when they note the behavioral impact of a terrorist attack, in particular the manner in which a person's behavior can change. For example, consider a person who becomes habituated to, say, stopping for coffee and pastries at a particular coffee shop every morning, and then the coffee shop is unavailable because it is within the Hot Zone. The person will find a new coffee shop to visit each morning; if this continues long enough then the person will develop a new habit that they are likely to continue even after access to their former coffee shop is restored and it reopens; to this can be added places to which the person went for lunch, coffee, dinner, and daytime shopping. The same amount of money might be going into the local economy, but the places that are forced to close until remediation is completed and access restored will have lost many of their former regular customers and developing a new customer base (and possibly luring former regular customers back) will require time. In this time, many of these businesses can be expected to fail with a concomitant loss of jobs and tax revenue for the city. This can also lead to a cycle of business startup followed by business failure that can last for many years.

This same effect can come into play with regards to tourism, conferences and conventions, and business travelers. Consider the impact of a radiological attack against the Las Vegas Strip. With the relaxation of gambling laws in the United States people can go to Atlantic City, casinos on Native American reservations, and any number of riverboats, cruise ships, and other venues to gamble. One can imagine them substituting any of these alternative destinations during the years required to remediate the Las Vegas Strip and then simply continuing to go to their new casino(s) even after Las Vegas is restored to the pre-attack normal. In addition to that, there might be a reluctance to visit a city that has suffered such an attack, whether worried about the risk of a second attack or the risk from residual radioactivity. This, too, can cause a lasting impact on the economy of the city that was attacked.

References

1. US Department of Transportation (2020) Traffic safety fact: Preview of motor vehicle traffic fatalities in 2019
2. Drottz-Sjöberg B, Persson L (1993) Public Reaction to radiation: Fear, Anxiety, or Phobia?. Health Phys 64(3):223–231
3. Stallings R (1984) Evacuation Behavior at Three Mile Island. Int J Mass Emergencies Disaster 2(1):11–26
4. Johnson G (2015) When Radiation Isn't the real risk. New York Times
5. Becker S (2004) Emergency communication and information issues in terrorist events involving radioactive materials. Biosecur Bioterror 2(3):195–207
6. Becker S (2005). Addressing the psychosocial and communications challenges posed by radiological/nuclear terrorism: key developments since NCRP report No. 138. Health Phys 89(5):521–530
7. Becker S (2011) Risk communication and radiological/nuclear terrorism: a strategic view. Health Phys 101(5):551–558
8. Becker S (2011) Learning from the 2011 great east Japan disaster: insights from a special radiological emergency assistance mission. Biosecurity Bioterrorism: Biodefense Strategy, Pract Sci 9, No. 4 Special Report
9. Becker S (2013) The fukushima dai-ichi accident: Additional lessons from a radiological emergency assistance mission. Health Phys 105(5):455–461
10. New York City Department of Health and Mental Hygiene (2018) Community reception center field guide. New York City
11. US Department of Homeland Security Science and Technology Directorate (2017) Radiological dispersal device (RDD) response guidance: planning for the first 100 minutes. US DHS, New York City
12. Hall E, Giaccia A (2019) Radiobiology for the radiologist, 8th edn. Lippincott Williams and Wilkins, New York

13. Preston D, Pierce D, Shimizu Y, Cullings H, Fujita S, Funamoto S, Kodama K (2014) Effect of recent changes in atomic bomb survivor dosimetry on cancer mortality risk estimates. Radiat Res 162(4):377–389
14. United Nations Scientific Committee on the Effects of Atomic Radiation (2010) Sources and effects of ionizing radiation: report to the general assembly with scientific annexes. United Nations, New York City
15. Centers for Disease Control and Prevention (2019) Radiation and pregnancy: Information for clinicians. https://emergency.cdc.gov/radiation/prenatal.asp, Accessed 14 Nov 2020
16. Karam P (2000) Determining and reporting fetal radiation exposure from diagnostic radiation. Health Phys 79(Supplement 2):S85–S90
17. Ketchum L (1987) Lessons of chernobyl: SNM members try to decontaminate world threatened by fallout. Soc Nucl Med (SNM) Newsline 28(6):933–940
18. Ryan C, Hall M, Richard C, Hall M, Chapman M (2006) Medical and psychiartric casualties caused by conventional and radiological (dirty) bombs. Gen Hosp Psychiatry 28:242–248
19. Zimmerman P, Loeb C (2004) Dirty bombs: The threat revisited. Defense Horizons 38(1):1–12
20. Karam P (1999) To remediate or not: A case study of Co-60 contamination at the Southerly wastewater treatment plant, Cleveland Ohio, USA. In Kjell Andersson (Ed) *Proceedings of the Values in Decisions on Risk (VALDOR) Conference.* Stockholm, June 13-17, pp 173–181
21. Hinds K (2012) Totaling sandy losses: how New York's MTA Got to $5 Billion. WNYC, Transportation Nation
22. US Environmental Protection Agency (1993) Federal guidance report #12: external exposure to radionuclides in air, water, and soil. Governmental Printing Office, Washington DC
23. Becker GS, Rubinstein Y (2011) Fear and the response to terrorism: an economic analysis. Center for Economic Performance Discussion Paper No 1079
24. Lester L, Wagner R, Saldana L (1997) Exposure of the pregnant patient to diagnostic radiations: a guide to medical management, 2nd edn. Medical Physics Publishing, Madison WI
25. New York City (2010) Radiological response and recovery plan
26. Radiation Effects Research Foundation (2020) Downloadable data, DS02 risk estimation: Solid cancer and leukemia mortality data. https://www.rerf.or.jp/en/library/data-en/ds02/. Accessed 14 Nov 2020
27. World Health Organization (2005) Chernobyl: the true scale of the accident. https://www.who.int/news/item/05–09-2005-chernobyl-the-true-scale-of-the-accident#:~:text=According%20to%20the%20Forum's%20report,danger%20to%20their%20health%20from. Accessed 14 Nov 2020

Chapter 9
How Nuclear Weapons Work

In a 1998 interview with Time Magazine's Rahimullah Yusufzai [1], Osama bin Laden stated that "Acquiring [WMD] for the defense of Muslims is a religious duty." In the few years prior to making this statement bin Laden had actively pursued nuclear capability, and his interest in nuclear weapons continued after this statement [2]. Nine years later, physicist Peter Zimmerman wrote an article for Foreign Affairs in which he described how a terrorist group that had been supplied with highly enriched uranium could build a nuclear weapon in a rented suburban home [3] just a few years after a similar account written by Brown et al (2006) appeared in a more academic publication. Zimmerman was not the first to suggest that terrorists might be able to build a nuclear weapon; this topic was explored in 1986, and likely earlier, by scientists with experience in this arena (Mark et al 1986) and a great deal of relevant information is available in the unclassified literature (e.g. [4], Reed 2007, 2009 and 2016; Brown et al 2006).

In the first two decades of this century there have been repeated attempted to smuggle and to sell highly enriched uranium to ostensible terrorists [5]. The fact is that terrorists have made it clear that they have been, and continue to be interested in obtaining nuclear weapons, and as Zimmerman and Lewis pointed out, if a group can obtain the weapons-grade uranium, they have achieved the hardest part of building a nuclear weapon. Given what we know about the destructive power of these weapons, it is reasonable to work to prevent an attack, and to develop plans for response as a city and as a nation should they be used.

9.1 Overview

A single nuclear fission releases about 200 million electron volts (MeV) of energy. By comparison, chemical reactions such as those involved in chemical explosions release on the order of tens of electron volts per reaction; thus, nuclear fission releases

P. A. Karam, *Radiological and Nuclear Terrorism*,
Advanced Sciences and Technologies for Security Applications,
https://doi.org/10.1007/978-3-030-69162-2_9

about a million times as much energy per kg of material when we account for the differences in atomic and molecular mass.

The release of so much energy in so short a period of time makes nuclear weapons particularly devastating and a nuclear attack in a major city could kill hundreds of thousands of people and destroy buildings over a large area; in Hiroshima and Nagasaki a single weapon of, by today's standards, relatively low yield destroyed the better part of two mid-sized cities, killing or injuring two thirds of the inhabitants. While state-developed nuclear weapons today are much more powerful than the bombs dropped on Japan in 1945, it is not unreasonable to believe that a nuclear weapon developed by a terrorist group could be comparable in yield to the devices used in Japan.

9.2 How Nuclear Weapons Work

Understanding how nuclear weapons work—in general, as the specifics are highly classified—is important if one is to understand why they are so devastating and why they can cause not only physical destruction, but radiological dangers as well.

9.2.1 Criticality, Critical Mass and Critical Geometry

Nuclear weapons depend on a runaway series of nuclear fissions to achieve their explosive yield. When a uranium atom fissions, it releases two or more neutrons, and if one of those neutrons (on average) goes on to cause another fission then the material is said to achieved criticality. Please note that "critical" in this usage is not intended to indicate risk or anything other than a configuration that permits a constant rate of fission over time. If more than one neutron (on average) goes on to cause a fission then the mass will be supercritical and power will increase; the rate at which power increases will depend on the average number of neutrons from fission that go on to cause secondary fissions and on the amount of time required for each generation of fissions. All else being the same, a larger mass of uranium (for example) will be able to capture more neutrons and will experience a more rapid increase in power.

No nuclear weapon will function unless it has a *critical mass* of material that is assembled in a *critical geometry*. Since neither of these factors are intuitively obvious it is necessary to explain what the terms mean.

Critical mass refers to the smallest mass of material that will sustain a nuclear chain reaction. This is based in part on the distance that a neutron from one fission must travel before it is able to cause a secondary fission; if the fuel is physically smaller than that distance then the neutron will escape from the mass of material before it can cause a fission and that neutron is lost from the weapon. The mass of material must be large enough so that enough neutrons from the majority of the fissions are able to go on to cause another fission before they escape.

But having a critical mass of material is not enough to ensure a chain reaction will occur. Picture a critical mass of plutonium that is hammered out into a flat sheet a few mm thick. In this configuration a neutron must be emitted within the plane of this sheet in order to cause a fission; if it is emitted at even a small angle it will likely escape the sheet before it can cause a fission. In this case there is a critical mass of material, but it is not in a critical geometry.

Now picture this sheet of plutonium being crumpled up like a sheet of paper. At some point it will reach a compact, nearly spherical configuration in which the neutrons are able to be captured before they escape—now we have a critical mass of plutonium that is in a critical geometry.

Thus, to construct a nuclear weapon one needs to have a critical mass of fissile material that can be assembled into a critical geometry (Note: there is a difference between the terms "fissionable" and "fissile." Fissile materials will undergo fission when struck by neutrons of any energy while fissionable materials require neutrons above a threshold energy level).

9.2.2 *The Fission Process*

If we begin with a single fission and assume that two neutrons from each fission go on to cause a secondary fission then in ten generations there will be over 1000 atoms fissioning, releasing about 32 mJ of energy. Another ten generations of doubling will increase the number of atoms fissioned by another factor of 1000, producing about 32 J. The next ten generations will increase the energy released to 32 kJ and ten more will bring this up to 32 MJ, about as much energy as about seven kg of TNT. Another ten generations will increase this yield to seven kt of TNT, the yield of a small nuclear weapon. Thus, 50 generations of fission with the number of fissions doubling each generation will bring the energy production from a level that can be measured only with specialized instruments to a point that will level a small city. Assuming a fission generation lasts about 0.01 μs [6], these 50 generations of doubling will take place in less than 1 μs.

The amount of energy released by a nuclear detonation depends on the number of fissions that take place so, in general, a larger amount of fissionable material will produce a larger number of fissions and a larger explosion. As noted above, a weapon must contain a critical mass of fuel in order to achieve a nuclear yield. A critical mass of fissionable uranium (^{235}U) is about 49 kg, and a critical mass of ^{239}Pu is about 17 kg as bare metal, although these masses can be reduced using various techniques [7].

With about 200 million electron volts (MeV) of energy released for each fission, the fission of every atom in a critical mass of ^{235}U (about 5×10^{25} atoms) would release the equivalent of nearly 400 kilotons (kt) of TNT (one kt is the amount of energy released by the detonation of one ton of TNT and is equal to about 3.8×10^{15}J). In reality, this amount of uranium is more likely to release about 10–20 kt of energy, because a large fraction of the atoms will not fission—the energy released at the start of the fission process adds so much heat to the weapon that it vaporizes, expands, and becomes too diffuse to sustain a chain reaction before more than a fraction of the uranium is fissioned. This release of energy also creates a shock wave that travels through the atmosphere, damaging buildings and people in the manner described in Chap. 7. This, and other physical effects, will be discussed in greater detail in the following chapter (Fig. 9.1).

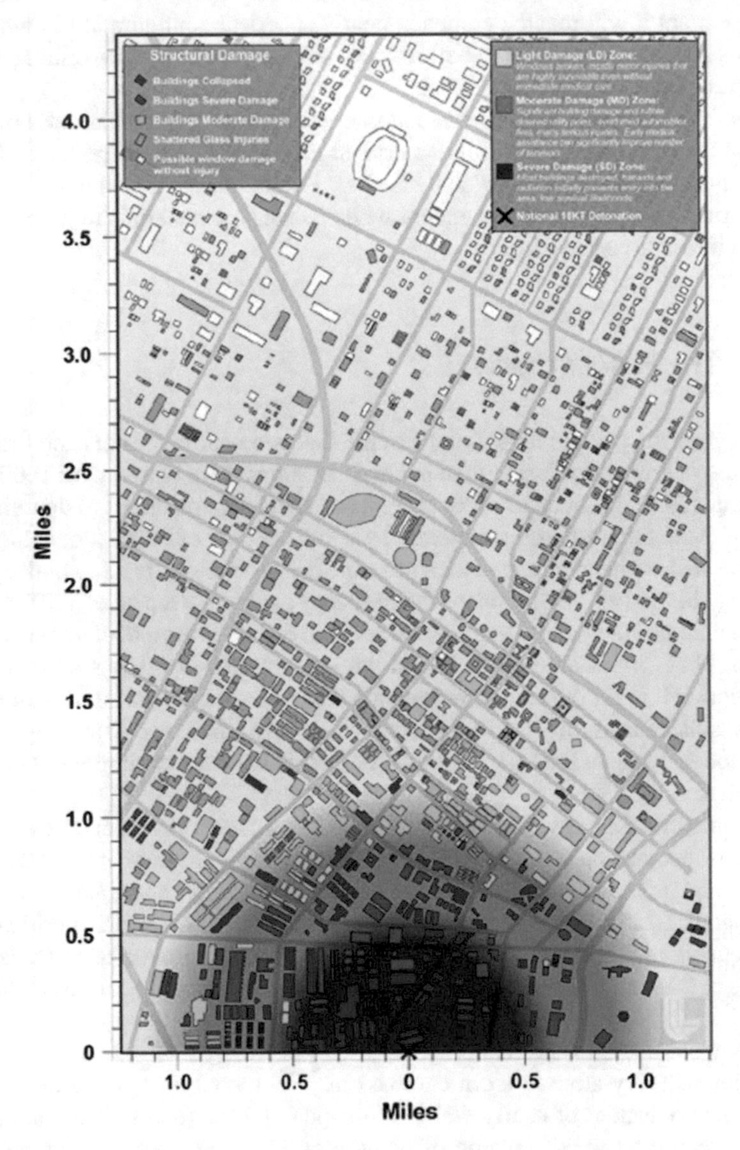

Fig. 9.1 Extent of damage caused by a 10 kT nuclear weapon in a notional urban environment [11]

9.3 Types of Nuclear Weapons

There are two fundamental types of fission weapons (we will not discuss thermonuclear fusion weapons as their fabrication is likely beyond the capabilities of a terrorist group), "gun-type" and implosion weapons. While both designs have been shown to function as designed, the gun-type device is less complex and is easier to design and build than an implosion weapon—it is simple enough that the US did not feel it necessary to even test the gun-type device that was the first nuclear weapon used. However, for reasons to be discussed below, plutonium cannot be used in a gun-type nuclear weapon; an implosion design is required for bombs using plutonium as the fuel.

9.3.1 Gun-Type Nuclear Weapons

In Zimmerman's nuclear terrorism article in Foreign Affairs he described how a terrorist group might construct a gun-type nuclear weapon. The reason for this is that such a weapon is simple to design and to construct; the major difficulty in constructing such a device lies with obtaining enough weapons-grade uranium to make it work. This is not to understate the difficulty of making a working nuclear weapon, but the fact is that a group that obtains enough highly enriched uranium can likely follow relatively simple instructions to produce a working nuclear weapon. It must also be noted that even a "fizzle" (a weapon that does not produce its designed yield) can produce a blast equivalent to several hundred Oklahoma City-style weapons, enough to destroy several city blocks and producing high levels of radioactive contamination.

In a gun-type device two pieces of weapons-grade uranium, each less than a critical mass, are slammed together at high velocity using an explosive charge. In one design a roughly cylindrical piece (the "bullet") is propelled down a large-diameter gun barrel into a cup-shaped target at the end of several rings of highly enriched uranium. While the target, the bullet, and the rings are each less than a critical mass of uranium, when assembled they achieve both critical mass and a critical geometry, a chain reaction ensues with the concomitant production of energy, and the weapon explodes with a nuclear yield [4, 8] (Fig. 9.2).

The reason Zimmerman feels a terrorist group is likely to use a gun-type device is that they are conceptually simple and are relatively easy to construct compared to an implosion-type weapon. The most significant drawback to gun-type devices is that they cannot use ^{239}Pu, a more effective nuclear explosive, but this is felt to be outweighed by the simplicity of their design.

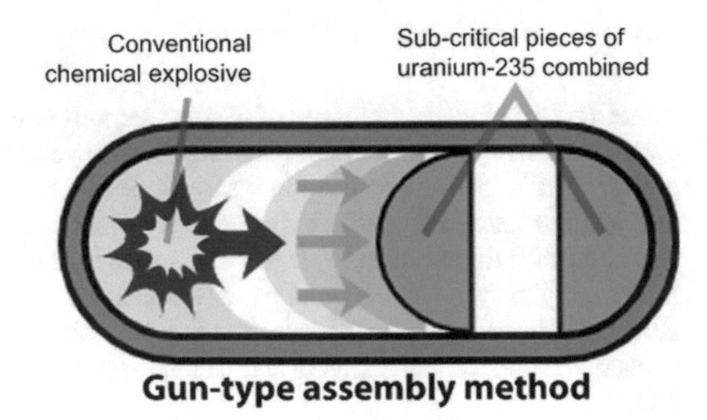

Fig. 9.2 Simplified diagram of a gun-type nuclear weapon (https://en.wikipedia.org/wiki/File:Fission_bomb_assembly_methods.svg)

9.3.2 *Implosion Weapons*

Plutonium is produced in nuclear reactors when ^{238}U captures a neutron to become ^{239}U and then emits two beta particles to become ^{239}Pu; the spent fuel can then be chemically processed to separate the plutonium from the uranium and the fission products. Although ^{239}Pu fissions readily, it can also capture a neutron to become ^{240}Pu. Normally this would simply be an interesting fact about plutonium, except for the fact that ^{240}Pu can undergo spontaneous fission, emitting neutrons in the process. These neutrons from spontaneous fission, if captured by the by ^{239}Pu atoms that comprise the majority of the weapon's fuel, can induce fission; in a gun-type device, this can produce enough energy to blow the weapon apart before a full fission yield is achieved. This is why ^{239}Pu is unsuitable for use in a gun-type device—because the two masses of fissionable material are brought together too slowly to prevent this pre-detonation from occurring [9] (Fig. 9.3).

Because of this, using plutonium in a nuclear weapon requires a different design; instead of assembling a critical mass from two subcritical pieces of material the

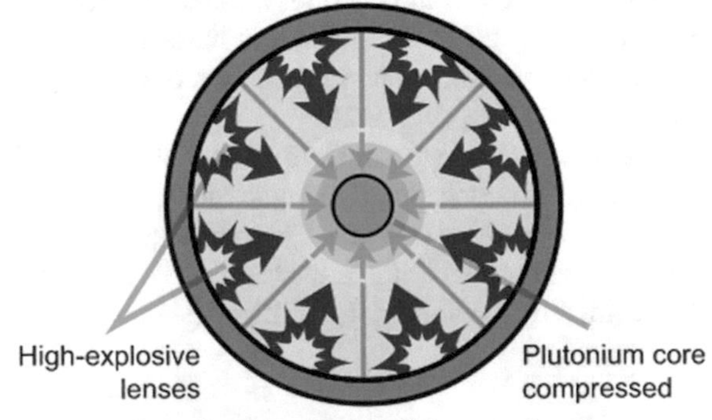

Fig. 9.3 An implosion-type nuclear device, such as the "Fat Man" bomb dropped on Nagasaki (https://en.wikipedia.org/wiki/File:Fission_bomb_assembly_methods.svg)

weapons designers construct a hollow sphere comprised of a critical mass of plutonium that is in a subcritical configuration. They then surround this sphere with explosive "lenses" that, when detonated, produce a spherically symmetric shock wave that compresses the sphere into a critical geometry, causing a nuclear chain reaction and a nuclear explosion.

The reason that this design is felt to be less likely to be used by a terrorist group is that it requires a great deal more skill to develop. In fact, during the Manhattan Project a great deal of research and testing was required to develop the explosive lenses that would produce a symmetric implosion. The fact that there are so many complex factors that must work nearly perfectly is why Zimmerman feels that a terrorist group is unlikely to fabricate its own implosion-style device.

9.4 Immediate Physical Effects

All of the energy produced by a nuclear weapon will be produced and released during the very short time, less than a microsecond, that the fission process will last. The majority of the energy released will be in the form of [6]:

- The kinetic energy of the fission fragment (about 82.5%)
- Gamma radiation from fission (3.5%),
- The kinetic energy of fission neutrons (2.5%),
- The kinetic energy of beta particles from fission products (3.5%),
- Delayed gamma radiation from fission products (3%), and
- Neutrinos from fission products (5%).

The kinetic energy of the fission products will be deposited within the weapon itself and within the rapidly expanding cloud of vapor formed during the detonation; this will heat the materials to incandescence and much will be re-emitted as visible and thermal radiation. The absorption of this energy by the air will cause it to expand rapidly, forming a blast wave that will move outwards at high speeds. At the same time, the visible light and thermal radiation escaping from the outer surface of the expanding fireball (the volume of air heated to incandescence by the heat of the explosion) will travel effectively instantly to be deposited in any objects in its way; if those objects are flammable then this can cause them to burst into flame, even if they are behind a window in a building's interior. Thus, fires can be ignited in areas distant from the explosion itself.

As noted above the air will initially heat up, causing it to expand rapidly enough to form a blast wave. At the same time, the heated air will begin to rise, leaving a partial vacuum at and near ground level; this will pull air in from all sides and will form the "stem" of the mushroom cloud associated with nuclear explosions. Although the explosion itself will be over within a second, the fires ignited by the pulse of thermal radiation will continue heating the air which will rise, pulling still more air in at the base from all directions. This air will help to fan the flames of the fires that were ignited until they merge into a mass fire, which will be discussed in the following chapter.

The National Academies of Science [10] describe three time domains in which the physical effects of a nuclear weapon detonation occur:

- *"The early time (less than 20 s).* In this first phase in the fireball, the mixing of ejecta and entrained sweep-up (dirt, vegetation, and rubble) with fission debris occurs.
- *The rise and stabilization phase (~10 s to 10 min).* In this second phase, the cloud rises and early fallout is produced. These effects are dependent on the local atmospheric profile, including temperature, relative humidity, and winds.
- *The late time (10 min to 2 days).* Wind transport and diffusion, as well as particle size, are major factors for the precipitation scavenging that occurs during this third phase."

References

1. Yusafzai R (1999) Conversation with terror (interview with Osama bin Laden). Time Magazine
2. Mowatt-Larssen R (2010) Al Qaeda weapons of mass destruction threat: Hype or reality?. Harvard Kennedy School Belfer Center for Science and International Affairs, Cambridge MA
3. Zimmerman P, Lewis J (2009) The bomb in the backyard. Foreign Policy, October 16, 2009
4. Serber R (1992) The los alamos primer: The first lectures on how to build an atomic bomb. University of California Press, Los Angeles
5. Downes R, Hobbs C, Salisbury D (2019) Combating nuclear smuggling? Exploring drivers and challenges to detecting nuclear and radiological materials at maritime facilities. The Nonproliferation Rev 26(1–2):83–104
6. Glasstone S, Dolan P (1977) The effects of nuclear weapons, 3rd edn. Government Printing Office, Washington DC
7. Paxton H, Pruvost N (1986) Critical dimensions of systems containing ^{235}U, ^{239}Pu, and ^{233}U. Los Alamos National Laboratory, Los Alamos NM
8. Rhodes R (1986) The making of the atomic bomb. Simon and Schuster, New York City
9. Reed B (2007). Arthur Compton's 1941 report on explosive fission of U-235: A look at the physics. American Journal of Physics 75(12):1065-1072
10. Reed B (2009). A brief primer on tamped fission-bomb cores. American Journal of Physics 77(8):730-733
11. Reed B (2010) Predetonation probability of a fission-bomb core. Am J Phys 78(8):804–808
12. Reed B (2016). A physicists guide to The Los Alamos Primer. Physica Scripta 91
13. National Academies Press (2005) Effects of Nuclear Earth-Penetrator and Other Weapons. National Academies of Science, Washington DC
14. US Federal Emergency Management Agency/National Security Staff (2010) Planning guidance for response to a nuclear detonation, 2nd edn. FEMA, Washington DC
15. Brown G, Carlyle M, Harney R, Skroch E, Wood K (2006). Anatomy of a project to produce a first nuclear weapon. Science and Global Security 14:163-182
16. Mark C, Taylor T, Eyster E, Maraman W, Wechsler J (1987) Can Terrorists Build Nuclear Weapons? Paper prepared for the International Task Force on the Prevention of Nuclear Terrorism

Chapter 10
Physical Effects of Nuclear Weapons

In the previous chapter we saw a description of the first seconds of a nuclear explosion; we will now see what happens in the minutes and hours that follow. In particular, we will look at the impact of the blast, thermal, and the radiological aftermath of the attack. At any given location these effects will vary according to the strength of the device and the distance from the point of detonation.

Yet another factor affecting nuclear weapons effects is the altitude of the explosion. Overpressure from surface burst (when a nuclear weapon is detonated at or very near ground level) is about 60% that of an airburst detonated at 1700 feet above ground, while the thermal effects of a surface burst are about 75% as severe as those from a burst at this altitude [1]. This is why the Hiroshima and Nagasaki devices were detonated as airbursts—to maximize the destruction caused by blast and thermal effects.

Very significantly, the altitude of a detonation has a dramatic effect on the amount of fallout produced by a nuclear explosion. Most materials that are within the volume of the fireball will be vaporized. As the fireball expands and cools, these materials will begin to condense and, as they do so, the radioactive fission products become incorporated into the condensing particles and they fall back to earth as fallout. As the detonation altitude increases, the amount of material incorporated into the fireball drops, and the amount of fallout decreases; the fission products fall to Earth over a larger area and form a more diffuse fallout plume.

We must remember that, although there will be massive loss of life and destruction up to a distance of a few miles from a surface burst in the range of a few tens of kt (the size of the Hiroshima and Nagasaki weapons), most of our major cities are far larger than this. This means that we can expect that most of a major city will survive a nuclear attack. We must also remember that large amounts of radioactive fallout will be generated, and this will spread far beyond the range of destruction; once we progress beyond the radius of destruction we should expect to see wide-spread radioactive contamination, possibly to great distances from the site of the attack.

P. A. Karam, *Radiological and Nuclear Terrorism*,
Advanced Sciences and Technologies for Security Applications,
https://doi.org/10.1007/978-3-030-69162-2_10

Each of these effects (blast, thermal, and radiological) will be discussed for weapons detonated at the surface and in the air.

10.1 Characteristics of a Nuclear Explosion

At the moment of detonation a nuclear weapon heats the material surrounding it to temperatures high enough to vaporize most known materials and to heat the air and vapor to incandescence. This roughly spherical volume is the fireball and it can be from a few hundred meters to a few km in radius depending on the weapon's yield. The energy absorbed by the air causes the expansion of the air that powers the blast wave; this is about 40% of the energy from fission. The remaining 35–45% of the fission energy is emitted in the form of thermal radiation that is deposited at a distance from the detonation.

10.1.1 Blast Effects

A moment after the fission reaction occurs the energy released heats the bomb and the air immediately surrounding it to temperatures of several tens of millions of degrees. This temperature is hot enough to vaporize any known materials. The resulting expansion of the air and vapors causes pressures to momentarily reach as much as a few million atmospheres before the gas begins to expand [1]. It is this initial release of energy, the heat it deposits, and the initial pressure it causes that initiates the destructive blast wave that radiates out from the site of the explosion.

As the heated air expands it forms a shock wave that rushes outwards from the point of detonation at high speed. The shock wave is initially supersonic but drops to sonic speeds within several tens of seconds. Initially, there may be two shock waves, the primary shock from the detonation itself as well as a secondary shock that is reflected from the ground in the event of an air burst; these merge within a few km to form a consolidated shock wave called the Mach front that propagates outwards.

The strength of the blast wave depends on the distance from the fireball, dropping in strength as it grows in area. Thus, a shock that is able to collapse a masonry structure at a distance of one km and that can kill a person through blast injuries at a distance of three or four km will only be able to break glass at a distance of ten km.

The effects of this blast will be largely the same in nature, although much greater in magnitude than those noted in Chap. 7. Broken glass, for example, can occur as far away as 4 km from the site of the detonation of a 10 kt explosion, affecting more than one million people in a densely populated city such as New York City. Closer in, buildings would likely be completely destroyed to a distance of 300 m and would be severely damaged to a distance of 1 km away [1]. This would affect more than a half million people in a densely populated city.

10.1.2 Thermal Effects

The great majority of the energy of the fission is in the form of the kinetic energy of the fission products; these fission products slow down and stop rapidly, depositing their energy in the air in the immediate vicinity of the weapon and heating that air to high temperatures. Initially the air is hot enough to emit x-ray energy radiation; these x-rays are absorbed and are re-emitted as lower-energy photons which are absorbed and emitted once again at somewhat lower energies, reflecting the cooling of the fireball as it expands. When the photons drop to a wavelength to which the air is transparent the thermal radiation escapes the fireball and exposes the surrounding areas.

Thermal radiation travels at the speed of light and will reach its target long before the blast wave arrives. This is clearly seen in some videos on nuclear weapons effects from the 1950s in which a puff of smoke is seen as the thermal pulse burns the paint and outermost layer of wood in a frame house built for testing weapons effects, followed by the arrival of the blast wave that tears the building apart [2].

According to Eden [6], who wrote about the thermal effects of nuclear weapons, an energy flux of about 10 cal cm^{-2} is sufficient to ignite many flammable materials. This energy can be in the form of infrared, visible light, ultraviolet, or x-ray radiation; as the photons are absorbed they deposit their energy and it is this energy that raises the temperature of the material absorbing them. With sufficiently high energy deposition the temperature will rise high enough that the material will ignite to start a fire; in an urban environment there will be many of these small fires ignited in apartments, offices, stores, and any other location in which flammable material is within the line of sight of the fireball.

As these fires grow, they will begin to merge and when enough small fires coalesce and are consolidated into a single conflagration the resulting mass fire is nearly impossible to extinguish, in part fanned by the winds formed by the hot air rising away from the fires. A mass fire can produce temperatures high enough to warp and even partially melt steel train tracks and close to 100% of combustible materials will burn. Hiroshima and Nagasaki suffered mass fires following their nuclear bombings, as did Dresden, Tokyo, and other cities attacked with non-nuclear incendiary devices during this same war.

10.1.3 Radiation Effects

Gamma and neutron radiation are released during fission and, from a device with a yield of about 10 kt, they are fatal to a distance of about 1 km from the site of the detonation. Gamma radiation is absorbed by whatever happens to be in its path; neutron radiation will eventually be absorbed as well, but this process can cause the material absorbing the neutrons to become radioactive as well. The neutron radiation can be dangerous in and of itself, but the neutron activation products tend to be short-lived and are not dangerously radioactive. And, while the gamma radiation dose is quite high, the radius at which the gamma dose rate is potentially fatal is smaller than the radius at which the blast and thermal effects will be fatal. More important, especially for a surface burst, is the radiation emitted by radioactive fission products that settle to the ground.

When uranium or plutonium atoms fission they each produce two radioactive atoms called fission fragments. Assuming an average fission energy of 200 meV we can calculate that a 10 kt explosion results from the fission of about 10^{24} atoms and produces about 2×10^{24} radioactive fission product atoms. In a surface burst, these radioactive atoms will attach themselves to debris swept up into the mushroom cloud or to the vapors of refractory materials that begin to condense as they rise into the sky. These will fall to the ground, forming the "footprint" of a fallout plume. An airburst, on the other hand, will not involve surface materials and there will be nothing to condense and scavenge the fission products. In an airburst, the fission products will remain dispersed in the atmosphere, drifting downwind and settling more slowly due to the smaller particle size; this results in a more tenuous plume, lesser deposition, and lower radiation exposure to those exposed.

What is the gamma dose 1 km from a 10 kt explosion?

A single fission releases about 200 meV of energy and 10 kt is the equivalent of about 2.6×10^{26} meV. Thus, a 10 kt explosion involves the fission of about 10^{24} atoms and releases an average of 7 meV of gamma energy [1] per fission. Thus, the total energy of the gammas from fission is about 7×10^{24} meV.

This energy will be emitted isotropically, passing through the surface of a sphere with an area of $4\pi r^2$. A sphere with a radius of 1 km will have a surface area of $4 \times \pi \times (10^5 \text{ cm})^2$, or about 1.26×10^{11} cm^2. Dividing the total gamma energy emitted by the surface area of this sphere gives an energy fluence of 5.6×10^{13} meV cm^{-2}, or about 9 J cm^{-2}.

By definition, 1 Gy is equal to the deposition of 1 J of energy per kg of absorber so the gamma radiation dose from the detonation itself will be about 9 Gy at a distance of 1 km from the bomb at the time it detonates, assuming that all of the gamma radiation is absorbed in water (a reasonable approximation of human tissue). The dose at other distances can be calculated using the inverse square law.

This value is comparable to the conclusions of the Radiation Effects Research Foundation's Dosimetry Study, which concluded that the gamma dose rate at a distance of 1 km from the Hiroshima weapon's detonation point was about 10 Gy with a neutron dose that was lower by about a factor of 10 (Figs. 10.1 and 10.2).

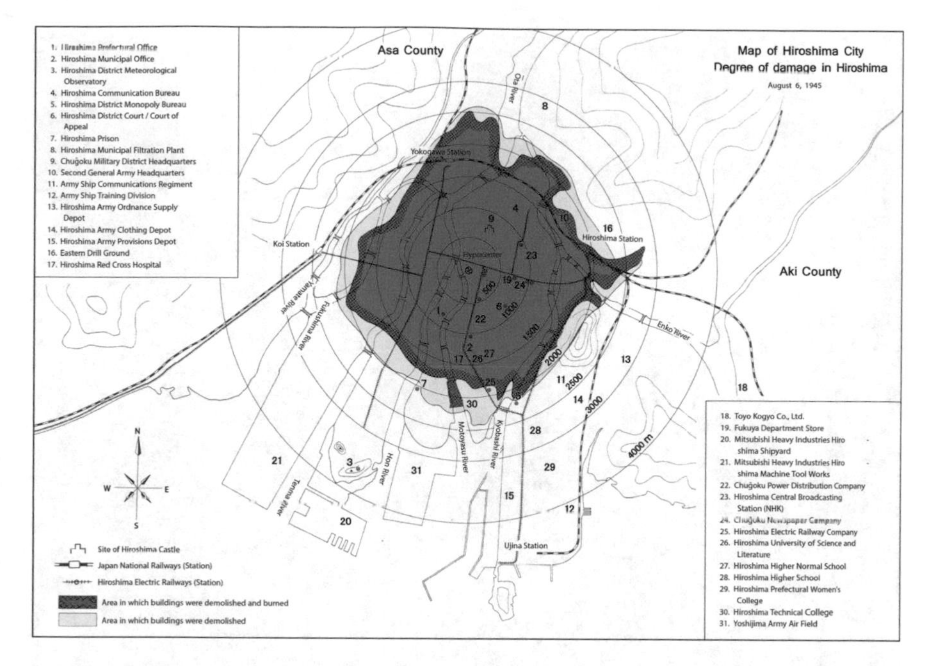

Fig. 10.1 A map of Hiroshima showing the radius of severe blast damage (yellow) and thermal damage (red). Courtesy of the Hiroshima peace culture foundation

10.1.4 Electromagnetic Pulse (EMP) Effects

The detonation of a nuclear weapon and subsequent ionization of the air and soil causes widespread ionization leading to a separation of positive and negative charges and causing electrical potential differences of up to one million volts per meter for a surface burst and several tens of thousands of volts per meter for an air burst [1]. This electrical current flow induces electrical current to flow in conductors—wires, cables, pipes, and so forth—as well as creating electrical and magnetic fields. All of this can damage or ruin electronic gear and high currents can trip circuit breakers or can start fires in wires (e.g. motor windings, household electrical wiring). Thus, a strong EMP can hinder radio communications, knock out computers and computer networks, and can damage city electrical equipment and distribution systems. The strength of the EMP varies with the altitude of the detonation, although its effective radius typically extends to distances only a few to a few tens of km from the site of the detonation, depending on the altitude and yield of the device used (Fig. 10.3).

Fig. 10.2 Map of fallout (orange and yellow) and blast effects from a simulated 10 kt attack. Note the bifurcated plume, due to upper-level and lower-level winds blowing in different directions [5]

10.2 Characteristics of Surface and Air Burst Detonations

As mentioned above, the initial high temperatures—hotter than the surface of most stars—cause the air in the vicinity of the weapon to become incandescent; this volume is the fireball. As the fireball expands it will cool, although the interior will remain hot enough to vaporize most materials for several seconds post-detonation. At this point, the fireball already contains the vaporized weapon casing and components, the fission products, and the air. If the fireball is in contact with the Earth's surface at this stage then it will also begin vaporizing the soil, concrete, buildings, and other materials with which it comes in contact. This is a surface burst. If the fireball does not come in contact with the Earth's surface then it is an air burst.

The altitude at which a nuclear weapon detonates can have a substantial impact on the physical and radiological impacts of the weapon. The weapons that were used on Hiroshima and Nagasaki in 1945 were set to detonate at altitudes of 580 m and 500 m respectively [1]. These altitudes were selected to maximize the blast effects

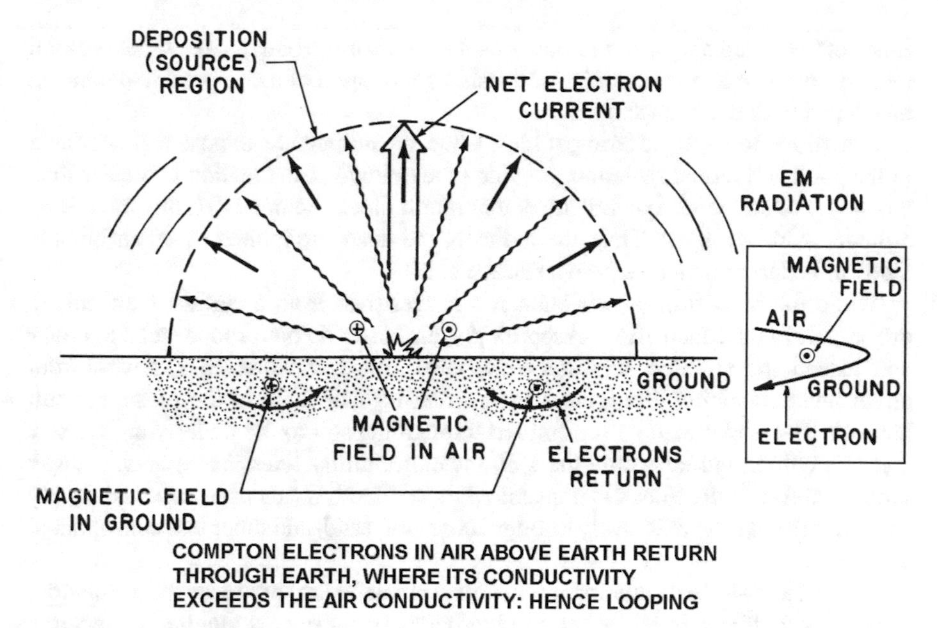

Fig. 10.3 Components of an electromagnetic pulse (EMP) [1]

of the weapons, and in each case the single weapon leveled the better part of the city as well as initiating mass fires.

A terrorist weapon, in comparison, is expected to be set off at or near the Earth's surface. While it is possible that a terrorist group might rent or purchase an airplane, that would add a layer of complexity to a likely already-complicated plot. Simply using a nuclear weapon in a terrorist attack will make a strong statement even without making the extra effort to achieve a modest increase in the damage radius. While a surface burst might not be quite as physically damaging as an airburst, it would have far more of a radiological impact than would an airburst; a surface burst can lead to fallout and contamination levels that would be quickly fatal to those who find themselves in the fallout plume. Given the significant difference in effects it makes sense to discuss the characteristics of surface and air bursts and how this affects their impact on the area attacked.

10.2.1 Air Burst

An air burst is a nuclear detonation in which the fireball does not touch the surface of the Earth. In an air burst the primary shock wave radiates outward from the detonation in all directions; when it strikes the ground it is reflected back upwards and outward. When this reflected shock wave merges with the primary shock wave the combined shock is stronger and more destructive than the primary shock alone—this is the

Mach effect noted above and it increases the force of the blast, making the weapon more destructive to a greater distance than for a weapon of a comparable detonated on contact with the surface.

In addition to having a stronger blast wave, an airburst can expose a greater area to the pulse of thermal radiation, provided the altitude of detonation is greater than the height of any terrain or structures that might block the line of sight from a low-altitude or surface burst. Thus, the radius of the mass fire ignited by an air burst is likely to be larger than that from a surface burst.

While the blast from an air burst is stronger than from a surface burst this is due in part to the Mach effect. Accordingly, air bursts do not produce a large crater and, indeed, most produce no crater whatsoever. However, when the blast wave from an air burst slams into the ground it can transmit that force to the rocks and soil beneath. This shock will be transmitted through the soil to the underlying bedrock and to anything buried within the soil, including utility lines, basements, subway tunnels, and so forth. Shock is transmitted most effectively through rock and tightly packed soil and least effectively through loose soil, sand, and other less consolidated materials.

While a high-altitude airburst (greater than about 30 km) produces the most widespread EMP effects, intermediate-level airbursts (from several hundred meters to a few tens of km) are the least effective at causing EMP [1]. Thus, an airburst intended to maximize blast and thermal effects will produce little EMP while one designed to maximize EMP effects will cause only minor blast and thermal damage.

10.2.2 Surface Burst

The fireball from a detonation on the ground is likely to be somewhat smaller than for an airburst as the ground will absorb much of the weapon's radiant energy [1]. In a surface burst, much of the fireball's thermal energy goes into vaporizing the ground and structures in close proximity to the site of the detonation and the remaining energy powers the blast wave and the thermal pulse. The blast wave will be less powerful than that from an air burst because, without a reflected shock, there is only the power of the primary shock to cause damage. These are all notable differences, but the most important differences are in other areas:

- EMP
- Cratering and shock transmitted through the ground
- Radioactive fallout.

The EMP effects of a surface burst are shorter-ranged but more intense than for an intermediate-altitude air burst. The shorter range is a result of the need to have a "line of sight" from the site of the detonation to the parts of the atmosphere that are ionized; the horizon (the part of the Earth's sphere that we can see) is closer at lower elevations, as are the parts of the atmosphere that can be exposed to gamma radiation traveling in a largely straight line. On the other hand, more material—the ground in particular—is ionized and is available to transmit the induced electrical current for a surface burst, and many conductors (buried pipes, utility lines, etc.) are buried in the ground or are in contact with it; thus the effects are more intense than for a blast at a greater distance from the ground [1].

The physical force of the blast wave will excavate a crater from the soil and rock beneath the weapon. Some of that material will be vaporized by the heat of the fireball and some will be thrown into the air by the blast. The precise dimensions will depend on variables that include the density and cohesion of the ground (e.g. sand, clay, shale, granite, etc.), the yield of the weapon, and its elevation or depth of burial. For a 10 kt nuclear device exploding on contact with the ground the crater will be about 75 m in diameter and 20 m in depth. This is deep enough to intersect with many underground utility lines and, in cities with subways, to do the same with train tunnels that are not buried too deeply [1, 3].

In addition to the cratering, the shock wave from the weapon will be transmitted through the rock and soil; denser and more-consolidated materials will transmit the shock more effectively than will softer sediments such as sand or loose soil. This shock wave can, in and of itself, sever utility lines, sewer and water mains, fire mains, and can collapse subway tunnels; it might also collapse basements and even deeper structures depending on their construction, distance, and the cross-sectional area presented to the blast wave. The precise effects of a surface burst on the underground infrastructure have not been published in the unclassified literature.

With regards to fallout, the vapors that had once been rock, soil, buildings, streets, and so forth will rise and cool and as they cool the vapors will begin to condense into a sandy substance with a fine to medium grain size [1, 4]. Some of these grains will incorporate radioactive fission products into them and more fission products will adhere to the grains' exterior.

The fallout plume will drift downwind, settling out as it travels. The heaviest and most radioactive fallout will tend to be found close to the site of the detonation and will decrease with distance. To a first approximation the fallout plume can be assumed to be Gaussian in nature with the highest concentration of fallout along the plume centerline and lowest at the edge. The tremendous thermal energy of the explosion and the subsequent fires will loft the plume to altitudes of a few to several km, and the winds at high elevations frequently blow in directions different than do surface winds. Thus, a nuclear explosion might well produce a complexly shaped plume with the major plume following the direction of the surface winds and the minor plume following the upper-level wind direction [5].

According to mathematical models the fallout from a 10 kt surface burst can produce dangerous radiation levels to distances of a few tens of km downwind with dose rates in the tens of Gy hr^{-1} in the centerline of the plume shortly after the explosion. Those who are in this area will be exposed to doses of radiation that can be harmful or fatal to a distance of 10 km downwind; in New York City this might include as many as one million residents, depending on the direction of the plume.

The amount of radioactivity present and the concomitant radiation dose rate will change with time because, as short-lived fission products decay to their longer-lived progeny (and longer-lived nuclides tend to have lower decay energies), the number

Table 10.1 The 7–10 rule showing radiation dose rate reduction versus time following a nuclear detonation

Time	Radiation dose rate (Gy hr^{-1})
1 h	10
7 h	1
49 h (~2 days)	0.1
343 h (~2 weeks)	10 mGy hr^{-1}
2401 h (~100 days)	1 mGy hr^{-1}

of atoms decaying each second becomes smaller, as does the energy associated with each decay. As a general rule of thumb, post-detonation radiation dose rates drop off according to the "7–10 rule" where, for every increase in time by a factor of seven the radiation dose rate drops by a factor of 10. For example, if we assume that the radiation dose rate in a particular part of the plume is 10 Gy hr^{-1} one hour post-detonation, subsequent dose rates will be as shown in Table 10.1.

It is also possible that a nuclear weapon might be detonated below ground; at a subway station, automobile tunnel, basement, parking garage, or similar location. Predicting the effects of a subterranean explosion is very difficult as there are so many unknown variables (e.g. depth of detonation, yield, pathways to the surface, etc.).

References

1. Glasstone S, Dolan P (1977) The effects of nuclear weapons, 3rd edn. Government Printing Office, Washington DC
2. Kuran P (1995) Trinity and beyond: the atomic bomb movie. visual concept entertainment
3. Vortman L (1977) Craters from surface explosions and energy dependence—a retrospective view. In: Roddy D, Pepin R, Merritt R (eds) Impact and explosion cratering. Pergammon Press, New York City
4. National Academies Press (2005) Effects of nuclear earth-penetrator and other weapons. National Academies of Science, Washington DC
5. Lawrence Livermore National Laboratory (April–May 2015) Helping cities prepare for a disaster. Science and technology review, pp 4–13
6. Eden L (2006) Whole world on fire: organizations, knowledge, and nuclear weapons devastation. Cornell University Press, Ithaca NY

Chapter 11
Health Effects of Nuclear Weapons

Even a small nuclear attack—a weapon that "fizzles" with a low yield—would be the most destructive terrorist attack in history; a fizzle would produce a blast at least an order of magnitude greater than that of the Oklahoma City bombing with all of the attendant physical effects of so powerful a blast (Reed 2011). To this would be added the thermal and radiological effects described in the previous chapter. And in addition to these we must also consider the long-term risk arising from radiation-induced cancer that might start appearing a decade and longer after the attack. A fully functioning nuclear weapon, even one with a yield similar to that of the weapons used in the Second World War, would have a far greater effect.

In an attack of such magnitude, however, the health impacts go far deeper than simple radiological injury, and they extend far beyond the injuries caused by conventional bombs. Dealing with burns alone from a single low-yield nuclear weapon is likely to exhaust any nation's capacity for care, and burns are only one of the myriad of non-radiological injuries that would be seen. In their modeling of the effects of an improvised nuclear device effects, Lawrence Livermore National Laboratories identifies several zones for a 10 kt nuclear weapon detonated at ground level [1], while Glasstone and Dolan [2] described the effects of electromagnetic pulse (EMP):

- Within about 300 m all buildings will be destroyed and 100% of those in this area will be killed
- Within about 1 km most buildings will be severely damaged and survival is unlikely for those in this area
- Within about 1–1.5 km buildings will be moderately damaged and the EMP will cause service disruptions and equipment damage that might be permanent
- Within about 4–5 km glass on exterior windows will be broken, injuring people struck by flying or falling glass
- Within about 10 km the EMP can cause temporary disruptions in electrical and electronic equipment

P. A. Karam, *Radiological and Nuclear Terrorism*,
Advanced Sciences and Technologies for Security Applications,
https://doi.org/10.1007/978-3-030-69162-2_11

- Within about 12 km the flash can cause temporary or permanent blindness; this can also lead to automobile accidents and injuries due to being suddenly unable to see
- Within a distance of about 20 km downwind of the detonation (depending on wind speed and direction) radiation dose to those in the fallout plume will sufficient to cause radiation sickness or death, depending on the protective actions taken by those in this area.

11.1 Prompt Fatalities

There are many who will be killed almost immediately by the blast, the temperatures, the radiation, by collapsing buildings and basements, in the mass fires ignited by the thermal pulse, and more. For a nominal 10 kt explosion this will likely include every person within a distance of 300 m and the majority of those within 1 km of the site of the detonation.

In addition to those killed in these zones, an unknown number might die in the seconds or minutes after a detonation when basements, subway tunnels, and other subterranean structures collapse from the shock transmitted through the soil and rock; others might die if they seek shelter in a location below the water table (or below sea level) that floods due to structural damage or the loss of dewatering pumps.

In a densely populated city such as New York City as many as 200,000 people might die in the first moments after an attack from the blast, heat, collapsing buildings, and falling glass and rubble.

11.2 Effects of Radiation Exposure in the Vicinity of the Detonation

As we saw in the previous chapter the gamma radiation dose from a 10 kt nuclear device will be approximately 9 Gy at a distance of 1 km from the detonation. Neutrons are emitted from fission as well, but calculating neutron exposure is more difficult because some neutrons are absorbed to cause fission, some are absorbed to cause radiation exposure, and all neutrons lose energy (and, hence, biological effectiveness) as they travel outward from the explosion. In fact, the matter of neutron dosimetry is sufficiently complex that it was not until 2003 that the matter was considered to have been resolved satisfactorily for the 1945 attack against Hiroshima [3]. Neutrons are thought to have accounted for only about 1–2% of the total absorbed dose for those within a distance of about 1 km, when we account for the higher relative biological effectiveness of fission-spectrum neutrons [4] their contribution rises to about 10–20% of the total biological dose [5].

For those who are directly exposed to the detonation with no buildings or terrain to shield them, the nearest distance at which they are likely to survive the radiation exposure is about 1400 m, at which distance the gamma radiation dose will be about 4 Sv and the neutron dose will be between 0.4 and 0.8 Sv. The cumulative dose, 4.4–4.8 Sv has a survival rate of about 50% in the absence of medical care and about 75% with expert medical attention [6]. This does not mean that the majority, or even half, of those at this distance will survive as radiation is only one of the hazards they face (the others being the blast, thermal effects, collapsing buildings, etc. mentioned earlier). Considering the large number of people (tens or hundreds of thousands, depending on the city attacked) many of these people are likely to be badly injured and unable to receive the medical care they will need in order to survive, thus the majority of these people are likely to die of their injuries and radiation exposure.

It should also be noted that the gamma radiation exposures mentioned above are for people who can see the location of the bomb—who are in the line of sight. The presence of terrain and of buildings, particularly large masonry structures, can reduce this exposure considerably, in some cases to survivable levels even for those only a few hundred meters from the bomb. However, the neutron exposure at these close distances might well remain dangerous because metal, glass, and masonry are not effective at attenuating neutron radiation.

11.3 Effects of Radiation Exposure in the Fallout Plume

The majority of those who are close enough to receive a fatal dose of radiation from the detonation of the weapon itself will not live long enough to die of radiation exposure; they will be killed by blast injuries, collapsing buildings, or the mass fires. And while the thought of 100,000–200,000 deaths in this zone is horrendous to contemplate, the potential death toll among those exposed to the fallout plume might be higher still. More importantly, while there is little that can be done to reduce the number of deaths from the explosion itself, taking proper actions can virtually eliminate all deaths among those exposed only to fallout.

A surface burst will vaporize many of the materials within the fireball and the heat of the explosion coupled with the fires will loft these materials high into the atmosphere where they will begin to cool and condense. This process will take some time to occur and the fallout will not begin to reach the surface again for a period of time following the detonation—the amount of time will depend on the amount of thermal energy and how high that takes material from the ground, the sizes of the particles that form, and the speed with which they drift back to the ground. During this time, the winds at both the surface and aloft will be carrying these particles downwind in whichever direction that happens to be at various altitudes; this will tend to scatter the fallout laterally (broadening the plume) as well as along the axis of its travel (lengthening the plume). These factors are to some extent random, which is why a Gaussian approximation will provide a reasonable first approximation of the distribution of radioactivity deposition across the plume. Along the plume's axis, however, the process is more systematic in that the largest and heaviest particles will tend to settle first, closest to the site of the detonation and the lightest particles will travel the furthest downwind. Thus, the concentration of radioactivity is expected to drop steadily along the downwind axis of the plume as one travels away from the site of the explosion. If there are multiple plumes, this same process will occur in each; if the winds are variable and changing direction frequently then this will help to further scatter and diminish contamination by the time it reaches the ground.

Radiation dose rates along the centerline of the plume might be as high as a few to several tens of Gy hr^{-1} initially, a km or so from the explosion, dropping to a few Gy hr^{-1} several km downwind and to tens or hundreds of mGy hr^{-1} even further away. All of these dose rates will drop rapidly with time according to the 7–10 Rule discussed in Chap. 9. Although radiation dose rates in the plume will be dangerously high in the first few hours following detonation, within one or two days they will drop to the point at which it will likely be safe to evacuate.

Knowing the amount of radioactivity produced per kt of yield, the amount of material produced by a blast of a given yield, the distribution of particle sizes (and subsequent distance traveled), and other relevant factors it is possible to determine the area that can be contaminated to produce dose rates of various orders of magnitude. According to a 2005 report by the National Academies of Science [7] "The area covered by the associated dose that would cause at least a 50 percent probability of fatality is roughly 2.6 square kilometers per kiloton, assuming that people are in the open and exposed for just the first day after the burst." Thus, we can estimate that 26 km^2 would be contaminated to this level in the aftermath of the 10 kt detonation we have been assuming, with larger areas contaminated to lesser levels.

These high radiation levels make it important for everybody in the city attacked, including emergency responders, to find shelter immediately and to remain sheltered until they can verify that it is safe to go outdoors. In the case of emergency responders this can be determined with their own radiation instruments; members of the public should shelter until they are told that conditions are safe in their location—this information can be delivered in person by emergency response or public health personnel or this information can be conveyed by television, radio, or internet. Other immediate response actions will be discussed in detail in Chaps. 15–19.

11.4 Non-radiological Risks in the Vicinity of the Detonation

While the radiological health effects are the ones that worry the public the most, the non-radiological health effects can be equally significant—and more so in locations outside the radius of the mass fires and structural collapse.

11.4.1 Burns

Among the most significant of non-radiological health effects will be burns caused by the thermal pulse or by the fires it causes. Japanese records indicate that between 50 and 75% of deaths among those who survived more than one hour following the blast were due to burns; it is likely that those who are severely burned will number in the thousands, depending on the nature and location of the blast. In Hiroshima the burn injuries included "mild erythema (reddening) to charring of the outermost layers of the skin" with flash burns noted among those who were as much as about 3–4 km away [2]. In some cases, severe burns were associated with the formation of keloids—thick scar tissue—that might have been caused by a combination of infection, malnutrition, and other factors associated with conditions that existed in Japan during and after the war.

Given the large amount of care required to treat patients with serious burns and the relatively limited number (about 1800) of dedicated beds for burn patients in the US [8] it may be difficult to provide adequate care for all burn patients. While these resources can be expanded somewhat, burn patients require a great deal of care, medication, pain relief, and specially trained staff to care for them properly; there is a limit to how far both personnel and materiel resources can be stretched and whether or not capabilities can be expanded sufficiently to accommodate all of the burn victims. These problems are likely to be exacerbated by the fact that many victims will not be able to survive transportation to hospitals with open burn beds (Fig. 11.1).

11.4.2 Blast, Broken Glass, and Crush Injuries

As discussed in Chap. 6, the blast wave itself will cause injuries to those exposed to it, in addition to those injured by broken and falling glass, falling objects knocked loose by the shock and blast, and buildings damaged or destroyed by the blast. These injuries might also include broken bones and amputations from people being thrown by the blast wave as it passes (called "displacement injuries"). Thus, medical centers and public health planners should anticipate the need to be able to treat tens to hundreds of thousands of persons with injuries severe enough to require medical attention, thousands of whom will be seriously injured and will require prompt attention. Injuries from falling objects and damaged buildings should extend to a radius of about 1–2 km and those from broken and falling glass can extend to a radius of up to 4–5 km from a 10 kt nuclear weapon.

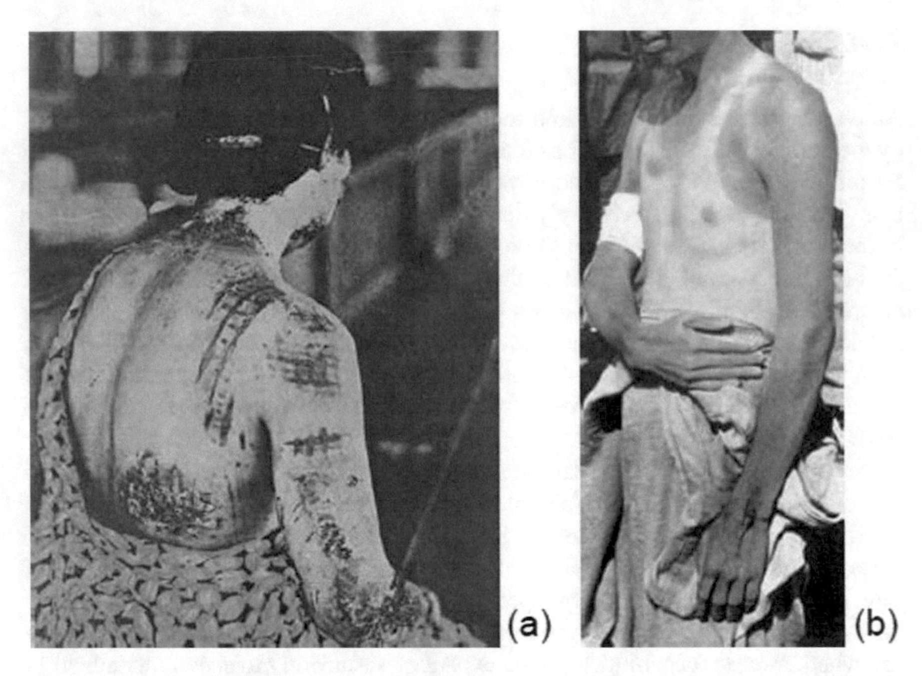

Fig. 11.1 Two Hiroshima burn victims, showing differential absorption of thermal energy by light versus dark (or no) clothing. US Government photos

11.4.3 Flash Blindness and Related Injuries

The flash from a nuclear explosion is tremendously bright—one early book about the American and German nuclear weapons programs was titled *Brighter than a Thousand Suns* [14]—and anyone looking in the direction of the bomb at the moment it goes off and the first few seconds afterwards can be blinded by the flash. While this flash blindness is likely to be only temporary for most, for some the deposition of so much energy on the retina can cause lasting damage to the tissue and can cause permanent blindness.

In addition to the flash blindness itself, being suddenly unable to see can lead to traffic accidents, it can blind pilots of aircraft of all sorts, there can be accidents at construction sites, even pedestrians who are unaware of obstacles in their path or oncoming vehicles. The number of people thus affected will depend strongly on the location of the bomb and the number of people with a direct line of sight at the time of detonation. With the population density of a large and densely packed city such as New York City, more than two million people can be close enough to possibly be blinded—more if the explosion occurs at night (Fig. 11.2).

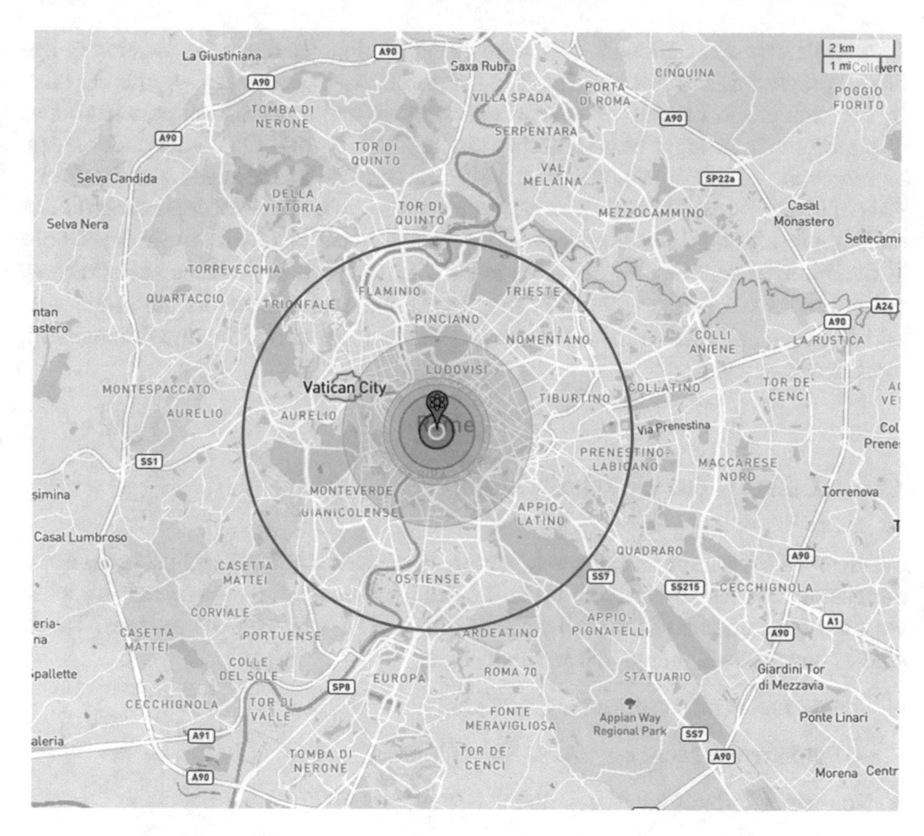

Fig. 11.2 Areas of various nuclear weapons effects; the heavy, medium, and light damage zones as well as the flash blindness area (blue circle). Generated using NukeMap (https://nuclearsecrecy.com/nukemap/) by Alex Wellerstein. Accessed December 9, 2020

11.4.4 Food, Water, and Electrical Shortages

As we saw earlier the shock from a nuclear weapon is likely to sever underground utility lines, including water mains. This can impair firefighting efforts in the short term and it will deprive citizens of water in the longer term. This can also deprive hospitals of water, hindering the quality of medical care they can provide until water can be restored to the city.

In addition, the shock and/or EMP might knock the city's electrical grid offline and restoring power might take days or weeks depending on the nature of damage. In addition to depriving hospitals and emergency response facilities of electricity, this will also impair the ability to store refrigerated and frozen food in homes and in stores, it can halt the pumps that provide water to high-rise apartment and office buildings, and can deprive the public of heating, air conditioning, and (for many) the ability to cook.

As if that were not enough, the nature of the incident itself and the location of the fallout plume might make it difficult to bring food and other supplies into the city. If, for example, if the plume were to contaminate major highways, port facilities, or airports then it might be impossible to feed the city until the contamination can be removed or decays to levels that would permit use of the roads and facilities. This might be a matter of days, or it might require several weeks depending on the radiation levels and the willingness of drivers and workers to work in various levels of radiation. In addition, drivers need to be willing to drive into these areas; during the author's work in the vicinity of the Fukushima nuclear reactors the group was told that food and other necessities were in short supply due to the unwillingness of truck drivers to enter the area.

11.5 Long-Term Considerations

Survivors of Hiroshima and Nagasaki have been tracked for nearly eight decades now; their health has been monitored, as has the health of their children to try to learn what we can of the effects of the nuclear attack on their health over the decades. These studies have shown, among other things, that:

- Some cancers increased among survivors
  - But have not been seen among their children
- Non-cancer effects have been noted as well, including
  - Cataracts
  - Thyroid tumors (benign)
  - Heart disease
  - Stroke
- Inflammatory reactions
- But no genetic effects among survivors' children.

A city that experiences a nuclear attack must also be prepared for these long-term impacts, especially induced cancers, if only because cancer is a disease that is both challenging to families and because of the cost of cancer treatment.

In addition, the nation that was attacked must be prepared to develop a registry and to commence long-term tracking of the survivors as soon as practicable, beginning with information that will make it possible to estimate the dose a person received (e.g. their exact location, biological indicators such as lymphocyte counts and lymphocyte depletion trajectories, chromosome aberrations, time until the onset of vomiting, and so forth). With time, it will be possible to refine these dose estimates and to look for correlations with the appearance of various symptoms and the time required for them to occur.

Maintaining this registry as an on-going project, as the Japanese and Americans have done with the Radiation Effects Research Foundation, will help to predict the health effects looming on the horizon so that they can be anticipated and planned for, but will also serve as an invaluable source of information to help us better understand the effects of radiation exposure over both the short and the long term.

A society should also be prepared to attend to the long-term mental health of those affected by a nuclear attack. Consider—a major city that suffers one million fatalities, a huge number by any measure, will also have several million survivors and many of those survivors will suffer from the trauma of the event for years or decades, not to mention commuters, tourists, and those living in proximity to the city that was attacked. Psychosocial issues have been shown (e.g. [9, 10, 11]) to be among the most common short- and long-term effects of radiological and nuclear disasters on the general public, as well as among those responding to these events. The society suffering such an attack must be prepared to address its citizens' needs in this area as well as their need for food, shelter, and short-term medical care.

References

1. Lawrence Livermore National Laboratory (2011) National capital region key response planning factors for the aftermath of nuclear terrorism (LLNL-TR-512111). Lawrence Livermore National Laboratory
2. Glasstone S, Dolan P (1977) The effects of nuclear weapons, 3rd edn. Government Printing Office, Washington DC
3. Straume T, Rugel G, Marchetti A et al (2003) Measuring fast neutrons in Hiroshima at distances relevant to atomic-bomb survivors. Nature 424:539–542
4. International Commission on Radiation Protection (2003) Relative biological effectiveness (RBE), quality factor (Q), and radiation weighting factor (w_R), ICRP Publication 92. Sage Publications, London
5. Wilson R (1962) A method for immediate detection of high level neutron exposure, by measurement of sodium-24 in humans. General Electric Co., Hanford Atomic Products Operation, Richland, WA
6. Hall E, Giaccia A (2019) Radiobiology for the radiologist, 8th edn. Lippincott Williams and Wilkins, New York
7. National Academies Press (2005) Effects of nuclear earth-penetrator and other weapons. National Academies Press, Washington DC
8. US Department of Health and Human Services, assistant secretary for preparedness and response, technical resources assistance center and information exchange (2016). Mass Burn Event Overview
9. World Health Organization (2005) Chernobyl: the true scale of the accident. https://www.who.int/news/item/05-09-2005-chernobyl-the-true-scale-of-the-accident#:~:text=According%20to%20the%20Forum's%20report,danger%20to%20their%20health%20from. Accessed 14 Nov 2020
10. Becker S (2011) Learning from the 2011 great east Japan disaster: insights from a special radiological emergency assistance mission. Biosecurity and bioterrorism: biodefense strategy, practice, and science vol. 9, No. 4 Special Report
11. Becker S (2013) The fukushima dai-ichi accident: additional lessons from a radiological emergency assistance mission. Health Phys 105(5):455–461
14. Jungk R (1970) Brighter than a thousand suns. Houghton Mifflin Harcourt, Boston
15. Reed B.C. (2011) Critical Mass, Efficiency, and Yield. In: The Physics of the Manhattan Project. Springer

Chapter 12
Societal Effects of a Nuclear Attack

The societal effects of even a small nuclear attack are likely to be profound, not only in the city that was attacked, but across society, possibly lasting for years or decades. While the long-term effects are beyond the scope of this work (in addition to being hard to predict with any degree of accuracy), we can draw upon the experiences of Hiroshima and Nagasaki to guess at the short-term societal effects of a nuclear attack today.

While a great deal of planning for response to nuclear attacks was conducted during the Cold War much of this planning is of only partial relevance to a nuclear terrorist attack. One reason is that Cold War weapons were typically high-yield devices with an average yield on the order of a megaton of TNT [17], while a terrorist device is likely to be less powerful by at least a few orders of magnitude. This, in turn, means that the thermal and blast effects will be far less powerful and less extensive than what Cold War planning anticipated.

In addition, Cold War nuclear arsenals were much larger than those of the present; at their peak in 1986 the US and USSR possessed 68,317 nuclear weapons between them with a further 1051 in the hands of the other nuclear powers [1]. Nuclear war planning assumed multiple high-yield detonations distributed across major cities as well as against military and industrial targets—the assumption was that there would be extensive destruction, heavy fallout across the city, and few (if any) unaffected locations. The experiences in Japan supported these assumptions, with a single relatively low-yield device visiting near-complete destruction on each city attacked.

A terrorist attack, however, is likely to consist of a single relatively low-yield device used against a large city with more robust construction than the largely wooden buildings of wartime Japan. Thus, from a societal standpoint, an act of nuclear terrorism is likely to be less destructive of the city attacked than Cold War planning or the experiences in Japan had led planners to expect; the former due to lower yields and fewer weapons used in an attack, and the latter due to having a larger city with more robust buildings.

P. A. Karam, *Radiological and Nuclear Terrorism*,
Advanced Sciences and Technologies for Security Applications,
https://doi.org/10.1007/978-3-030-69162-2_12

That being said, a terrorist attack's effects on the larger (national) society is likely to resemble our experiences in Japan more than to Cold War planning because it would most likely involve a small number of relatively low-yield nuclear weapons, leaving the majority of the nation intact and prepared to render assistance to the survivors.

12.1 Public Health

The effects of a nuclear detonation on the health of individuals was discussed in Chap. 11, here we will examine the public health effects.

12.1.1 Radiation Injury

The most obvious concern following a nuclear attack is the number of people expected to suffer from radiation illness and injury. Flynn and Goans [2] discuss four primary categories of patients, grouped by their radiation exposure and the expected symptoms. The following paragraphs are summarized from their paper.

- Group I patients are those with an exposure of less than about 2 Gy. These patients might or might not require hospitalization, depending on their symptoms and their lymphocyte count. In the short term, they should be kept under close observation with periodic complete blood cell counts (with differential) and can be managed as outpatients unless their physical injuries (e.g. thermal burns, lacerations) require attention. No patients in this group are likely to die of their radiation exposure alone, although the majority of those with exposures greater than 1 Gy are likely to fall ill with radiation sickness.
- Group II patients are those with exposures between about 2–5 Gy. These patients will require hospitalization and, without proper medical care, some of these patients may die of their radiation exposure. Radiation can kill lymphocytes and will impair the body's ability to produce new lymphocytes for several weeks or months, as a result reverse isolation will help protect against succumbing to opportunistic infections until the immune system can become reestablished. Early growth factor (cytokine) therapy and prophylaxes (viral and fungal), and antibiotics should be considered, along with selective gut decontamination using oral antibiotics.
- Group III patients are those who have received an exposure of between 5 and 10 Gy and all of these patients should be hospitalized with reverse isolation and early growth factor therapy. These patients, too, should receive selective gut decontamination as well as viral prophylaxis and anti-fungal therapy; antibiotics should be administered early in anticipation of severe neutropenia. Flynn and Goans also recommend that both Group II and III patients receive assistance from

a hematologist who has an understanding in this area to manage those aspects of treatment. Every Group III patient will develop severe radiation sickness, some may suffer from radiation burns, and a high percentage are likely to die from their radiation exposure, especially at the higher levels of exposure.
- Group IV patients will have received a dose in excess of 10 Gy and they will not survive their exposure. Thus, caring for Group IV patients should consist of supportive care and treating symptoms as they arise. If resources permit, these patients can be accorded additional care, up the level provided to Group III patients.

In addition to the whole-body effects of radiation exposure, many patients in Groups II, III, and IV might also develop radiation injury to the skin, although many of these effects do not appear for several weeks after exposure. These effects include erythema (skin burns) at a skin dose of approximately 3 Gy with dry desquamation (peeling) at a dose of about 10 Gy to the skin and wet desquamation (blistering) at 20 Gy to the skin. If this skin exposure is received from penetrating (gamma and neutron) radiation the patient will likely die of radiation sickness; if the skin exposure is the result of contamination with beta-emitting fallout then the patient might survive the exposure [3].

Patients in all exposure groups might experience nausea, vomiting, and diarrhea and these will worsen as the exposure increases. Thus, dehydration and electrolyte depletion should be expected and monitored, and appropriate antiemetics and/or anti-diarrheal medications considered. Flynn and Goans do not recommend any except for the most necessary surgery until the patient is past the worst of the radiation sickness.

For American cities and hospitals that lack the expertise and resources to address the number of radiation injury patients they are confronted with, the Radiation Injury Treatment Network can provide facilities and expertise, including transporting patients to member hospitals in other cities for care and treatment. In addition, the Radiation Emergency Assistance Center and Training Site (REAC/TS) can provide medical advice and assistance.

12.1.2 Burns, Lacerations, and Other Injuries

As noted in Chap. 11, those with non-radiological injuries are likely to be quite numerous. In particular, thermal burns from the detonation are likely to affect tens of thousands of people with burns from the mass fires injuring still more. It is likely that the number of serious burns will outstrip the ability of the city to care for burn victims and many patients—possibly the majority of burn victims—will need to be transported to a facility that can treat them properly or will have to be treated in ad hoc burn units.

In addition, blast injuries, lacerations from broken glass, broken bones, and other physical injuries are almost certain to overwhelm available resources; many with light injuries will likely need to go untreated initially, until additional medical resources arrive from elsewhere. On top of that, the flash blindness can cause temporarily blinded drivers to have accidents, leading to tens or hundreds of thousands of additional injuries (and patients).

Medical caregivers will almost certainly be required perform extensive triage and this might require delaying treatment to some of the first patients to arrive in favour of more badly injured patients that arrive later by ambulance. This process might continue for a few to several days post-blast as rescue and evacuation operations continue, as shelter-in-place orders are lifted, and as access is restored to different parts of the city, making it possible for more of the injured to reach medical facilities.

12.1.3 Medical Capabilities

It is possible that some medical facilities might be destroyed by the blast or made inaccessible by the fallout plume, or that emergency and medical responders might be unable to reach their places of work immediately following a nuclear attack if they are sheltering in place, if roads are destroyed or blocked, if the personnel die in the attack, or if they do not respond due to concerns for their safety or for the safety of their families [4, 5].

In addition to a lack of personnel, it is entirely possible that hospitals and other treatment facilities might exhaust their supplies, decreasing their ability to render medical assistance until they can be replenished. In the United States and other nations that maintain national stockpiles of medical supplies and medications these shortfalls should begin to be alleviated within a few days; nations that do not maintain such a capability might be slower to replenish their hospitals. We must also consider that medical personnel are likely to become exhausted, as was all too common in nations and cities struck by the 2020–2021 coronavirus pandemic, and this can affect judgement, the quality of care provided, and the number of patients that can be cared for.

Medical capabilities are also likely to be affected by the loss of electrical power and utilities in affected areas, along with the possible effects of the EMP on hospital computers, diagnostic equipment (e.g. CT, nuclear medicine cameras, MRI), and other electronic gear. These losses can have wide-ranging impacts on the ability of the hospital and its staff to provide medical care. They might also require staff to maintain patient charts in hard copy (as opposed to digital) formats and may make it difficult to obtain patient records to check for existing medications and medical conditions, let alone to track attack-related injuries and symptoms.

The net result of all of this is that a city's medical capabilities is likely to be severely reduced in the aftermath of a nuclear attack. This, in turn, is almost certain to affect not only those members of the public affected by the attack, but others as well who suffer the normal illness and injuries that occur on a regular basis.

12.1.4 *Food, Water, Shelter, Security*

In an April 2017 exercise in which the author was a participant, a (simulated) nuclear weapon was detonated near New York City. Over the course of a three-day multi-agency exercise that included personnel from a number of major NYC agencies and representatives from multiple local, state, and national agencies the participants spent the time in discussion, working through the major issues that would be posed by such an attack and how each of the participants would work to resolve them. Among the surprises noted by many participants were the difficulty of attending to so many issues that had little to do with radiation exposure—specifically, providing for food, water, shelter, and security for those in even those parts of the city not affected by the blast or the fallout plume [6]. In addition, given the likely need to evacuate large parts of the city, a great deal of time was spent in discussing the optimum time to evacuate areas within the fallout plume as well as which other parts of the city might require evacuation.

Among the problems discussed was that the blast itself and the accompanying EMP might knock out water supplies to much of the city; the crater and underground shock wave are likely to rupture a large number of water mains and the EMP can also damage the city's electrical grid as well as directly affecting the pumps needed to raise water to the top floors of high-rise buildings and to de-water subway tunnels. In addition, the loss of utilities can also interfere with the operation of airport and port facilities, leading to loss of ability to bring in supplies. This would be exacerbated by the possible loss (in the case of New York City) of major bridges and tunnels; even if they are not destroyed or collapsed by the blast itself, it would be unwise to use them until inspected and their structural integrity assured. Temporarily lacking the ability to bring in food and water (as well as medical and other supplies) by air, sea, or land would pose a grave short-term public health concern until access via one or more of these avenues is restored. Not only that, but the same possible loss of structural integrity that might limit the movement of supplies into the city could also limit the ability to evacuate the public. The George Washington Bridge, for example, has a main span that is 110 m in length and is about 36 m in width for a usable area of about 6600 m^2 for the bridge's two levels. If the bridge is packed with people standing about ½ m apart (four people per square meter) with people weighing an average of 90 kg apiece then the weight of people on the bridge's main span will be about 2.4 million kg or 2400 tons, the weight of about 1000 cars. The other major bridges from Manhattan to Brooklyn would suffer from similar loading it would be unwise to permit so many people—and so much weight—onto a bridge whose structural integrity was not certain; loading would not be a concern in any of the four major tunnels (Hudson, Lincoln, Battery, and Midtown), but we would not want them to develop leaks when filled with evacuees.

Those who cannot evacuate must shelter, preferably in their homes. Unfortunately, in many major cities (of which NYC is, perhaps, an extreme example), many residents do not necessarily maintain large stocks of food and drink in their homes, relying instead on food and grocery deliveries, food stands, restaurants, and frequent small trips to grocery stores. Thus, a city such as New York City can include hundreds of thousands or even millions of residents who lack more than a minimal food supply. In New York City, for example, the author visited a few hundred families in the aftermath of a major storm, noting that many were running short of food and water even just one week after the storm; the author saw something similar in the Fukushima area following the 2011 tsunami and nuclear accident. Not only that, but in the Fukushima area, although roads were open and not heavily contaminated, drivers were unwilling to make deliveries into contaminated areas, leaving those who did not evacuate with difficulties finding food for several weeks post-accident.

Providing shelter for residents will also pose a serious public health concern in the first days and weeks post-blast. Many in the moderate and light damage zones as well as those in areas suffering from thermal radiation effects will be forced from their homes due to blast damage and/or fires; others by radiation dose rates from fallout. Broken windows may make many dwellings unlivable (particularly during very hot, cold, or rainy weather), and in the absence of electrical power, apartments on high floors might be virtually unreachable by all but the most fit residents. Thus, there might be a need to provide shelter for tens or hundreds of thousands of people for the first few weeks, until utilities begin to be restored, food and water become more widely available, and radiation dose rates in parts of the fallout plume fall to acceptable levels.

Once shelters are established it will be necessary to minimize the spread of radioactive contamination to them. In Japanese shelters following the Fukushima accident this was accomplished by establishing a survey point at the entrance and all persons returning to the shelter were required to be surveyed prior to entry. It will also be necessary to ensure that shelters are provided with food, water, bedding, medical supplies, and recreational materials, as well as attending to the needs of the various populations within. The shelters in Fukushima Prefecture, for example, made an effort to place elderly and infirm residents in proximity to the medical station whenever possible, as well as stocking children's games, books for all ages, and even arranging for regular visits by celebrities and various performers (e.g. high school bands, dance groups, etc.) for residents who were sheltering for several weeks.

During the nuclear terrorism exercise mentioned earlier the issue of public security was also discussed. Law enforcement officials noted the possibility of looting and lawlessness in the immediate aftermath of a large-scale attack and suggested the potential need to ask residents to shelter in place until civil order can be restored, providing this will not cause excessive radiation exposure among those located in the fallout plume. This is another example in which safety involves many factors of which radiation is only one, and in which one must not only be aware of, but must be able to prioritize the entire panoply of risks posed by a complex environment (Fig. 12.1).

Another aspect of security to consider is the security of emergency and medical personnel. During a conversation with emergency room physicians from New York City's Saint Vincent's Hospital one of the staff physicians mentioned that, in the aftermath of the terrorist attacks on the World Trade Center, injured patients were lined up in the street in front of the hospital while staff performed triage. The physician

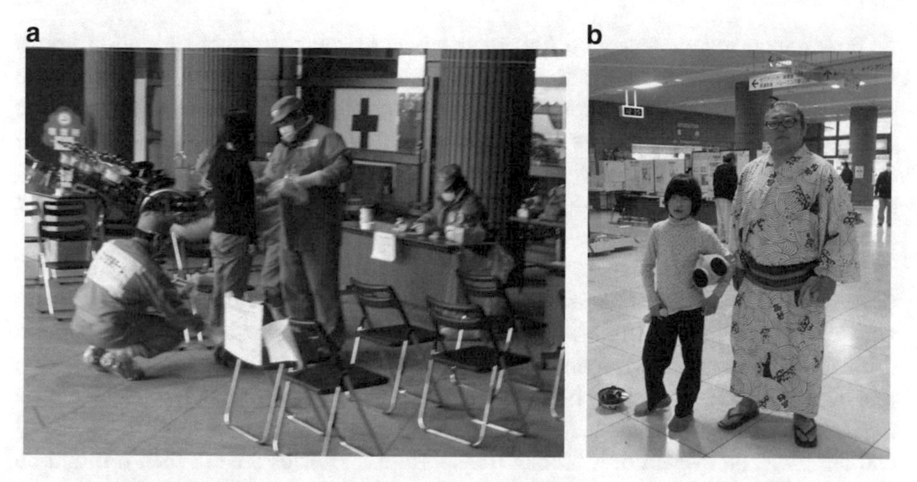

Fig. 12.1 **a** and **b**: Contamination survey of shelter resident prior to re-entering the shelter (L) and a visiting sumo wrestler posing with one of his fans inside the shelter. Author's photos

then shook her head and said "It never occurred to us that this made us a target for a secondary attack. It didn't happen then, thankfully, and we won't make that mistake again." Force protection any terrorism response plan; including searching for secondary explosives, snipers, and other threats to responders and the public both. This will be discussed in greater detail in Chap. 16.

12.2 Access Restrictions to Affected Parts of the City

The most obvious areas to which access will be restricted or impossible are those areas closest to the seat of the explosion. Mass fires will prevent entry into the areas less than 1 km (or so) from the bomb's location, radiation from the fallout will prevent safe access into the surface footprint of the fallout plume, and the presence of rubble (and the crater) will restrict access to the site of the explosion even after radiation dose rates decay to safe levels.

While the majority of the city attacked will be unaffected by the blast or fires, the fallout plume can make large areas inaccessible due to dangerous radiation levels and high levels of contamination; the plume footprint will be too dangerous to enter at first and the plume can present an obstacle in and of itself. The fallout from a nuclear device that explodes in Midtown Manhattan, for example, can extend across the island of Manhattan, effectively cutting the island in half and necessitating establishing access locations in both the north and the south parts of the island. If the plume extends further (and it might well stretch for 20–30 km downwind), areas on either side of the plume could be isolated from each other as well. At the same time, access to areas within the fallout plumes will also be restricted, at first due to radiation levels and later because of high levels of contamination.

Emergency responders who find themselves or their precinct or station house within the footprint of the plume will be unable to respond until radiation dose rates drop to safe levels; the responders can determine this themselves if they have appropriate radiation detectors and are trained in their use or they can shelter until they are told that it is safe to evacuate. Either way it will be important for them to communicate their status—including their health—periodically to their chain of command.

The effect of lack of access to these areas is difficult to predict since the areas to which access is restricted cannot be known in advance of an attack, thus there is no way to predict what might lie within those areas. For example, if these areas contain critical emergency management and communications facilities then the short-term response to the attack might be hampered at a time when organized leadership is most important. If these areas include major medical centers then the city's medical response will be impaired; if crucial transportation arteries are cut then it might be difficult or impossible to move people, responders, or supplies around the city. To this last point, one discussion that took place during the New York City IND exercise was on the topic of using fire boats to wash contamination from the deck of the George Washington Bridge and from the bridges across the East River to facilitate safe evacuations from Manhattan.

An important early task will be to delineate those areas that are unsafe to enter or in which entry is not possible due to physical damage to the city, and the manner in which these areas change with time. In fact, knowing the outlines of the radiological "no-go" areas will be an important part of each day's activities so that responders understand where they can (and cannot) operate as they attempt to stabilize affected parts of the city, to evacuate the public, and to extend the areas within which they can safely operate.

After the emergency phase ends it will be necessary to determine the criteria needed to restore access to and to reoccupy areas contaminated by fallout. These criteria—radiation dose rate, contamination levels, anticipated dose to residents, concerns about resuspended contamination, and so forth—are likely to be contentious, with many stakeholders demanding cleanup to pre-attack background radiation levels before considering reoccupancy and others recommending cleanup standards based on the risk posed by residual contaminants. A third option would involve limiting access to contaminated areas and letting them decay to background over the course of years or decades, and yet another option might combine aspects of all of these—remediating the most essential areas, restoring limited access to others, and restricting access to still other areas until they decay to pre-attack levels of radiation. Each of these possibilities represents an attempt to balance the radiological and other risks against the short- and long-term economic, health, political, and social impacts on a large number and variety of stakeholders, some of whom will understand the relevant radiation science and most of whom will not. Thus, restoring access to areas restricted on account of physical damage and contamination is likely to be a controversial process that might take years to resolve to the satisfaction of all stakeholders.

12.3 Psycho-Social

In 2005 the World Health Organization conducted a thorough examination of the health effects of the Chernobyl nuclear reactor accident [18]. Among their observations was that the long-term psychological effects of the accident on those living in the area were more significant than were the long-term health effects of exposure to radiation during and after the accident. The WHO noted elevated levels of depression, anxiety, substance abuse, and self-harm among not only those evacuated from contaminated areas, but also among their children, including children not born at the time of the accident.

In Japan, an entire generation of Hiroshima and Nagasaki survivors, the hibakusha were subject to bias and discrimination for decades following the attacks. In the words of American author Studs Terkel, who interviewed two hibakusha (atomic bomb survivors) while writing his book *The Good War:* "There is considerable discrimination in Japan against the hibakusha. It is frequently extended toward their children as well: socially as well as economically. *'Not only hibakusha, but their children, are refused employment,' says Mr. Kito. 'There are many among them who do not want it known that they are hibakusha.'*"(Terkel 2011) There is every reason to assume that similar stigmatization would affect the survivors of a nuclear attack in the US, Europe, or elsewhere.

The fact is that the detonation of a nuclear weapon will be traumatic for those in the city that was attacked and for the rest of the nation that was attacked, albeit to a lesser degree. The NCRP describes some of this:

> The shock of the attack will still be strong and dreadful images such as those seen after the Oklahoma City bombing and the September 11, 2001 attacks may be fresh in people's minds. Some affected individuals will have suffered the death of loved ones. Others may still be waiting for word about missing relatives. Families will be deeply worried about their children and women who are expecting babies will be anxious about their pregnancies. Depending on the situation, there may also be concerns about environmental contamination and the safety of food and water supplies. In addition, there may be a continuing threat and fear of possible additional attacks. In summary, people will be shaken and on edge. [7]

NCRP also notes that emergency responders will be torn between their inclination and desire to serve their community and their concerns about radiation, their individual and organizational ability to respond appropriately, and the potential effects on their health. In fact, this report cites a number of surveys in which both emergency and medical responders exhibited a greater reluctance to respond to a radiological or nuclear emergency than to natural disasters, an influenza epidemic, fires and structural collapse, or other "routine" disasters; their willingness to respond to a radiological or nuclear disaster was comparable to that of chemical and biological attack.

Emergency responders also face the threat of PTSD and other mental health issues during and after response to a disaster of this magnitude, in addition to difficulties arising from working long hours, lack of sleep, and stress [8]. Hall et al. [9] make similar observations, and also stress the importance of attending to the psychological care of both responders and the public, noting that new cases of psychological trauma can continue to manifest up to two years after terrorist attacks using conventional explosives, they also acknowledge that "Nuclear disasters have been *more likely* to induce chronic psychopathology than other types of disasters…" (emphasis added).

All of these authors emphasize the importance of communications in the immediate aftermath of any large terrorist attack, and particularly in the aftermath of a radiological or nuclear attack. Emergency responders need to know where to report to work, whether or not their stations are safe, and they need to be able to report to and receive direction from their chain of command. Some of this information is also vital for the general public, in particular the location of the footprint of the fallout plume, when it is safe to evacuate (and in which direction), as well as reminders of the need to shelter to avoid receiving a fatal radiation dose. Unfortunately, due to the effects of EMP, this might not be possible for some time after an attack, which might be too late to help those who do not know in advance that, as an immediate protective measure, they should shelter in place until they are told that it is safe to go outdoors and/or evacuate.

Among others, Becker [10–12] has noted that accidents involving radiation are typically communicated poorly to the public. In particular, information presented to the public can be conveyed by persons who are not felt to be trustworthy or who are not experts in the areas being communicated (e.g. engineers or administrators providing health advice) or the information and instructions provided appear to be inconsistent from day to day. In the aftermath of the Fukushima nuclear reactor accident the author was involved in a number of conversations with personnel at the local, state, and national levels in both the US and Japan and noted that many state and local governments were reluctant to release information or recommendations to the public until the national government had issued statements for fear of being contradicted, at the same time the national government was slow to do so owing to worries about making minor errors that might make the government look bad. Unfortunately, this meant that, in the absence of information from official sources, unofficial sources, including many who were not knowledgeable about the accident or about radiation, gained a prominence that would not otherwise have been the case.

Among the concerns raised by the NCRP [7] is that "*women who are expecting babies will be anxious about their pregnancies*," a concern that has been noted in many circumstances. In the aftermath of Chernobyl, for example, the USSR encouraged women exposed to radiation to terminate their pregnancies on account of the radiation exposure, and many women in non-USSR Europe were advised to have therapeutic abortions by their physicians. While the exact number of abortions is not known, estimates range from 100,000 to more than a quarter million [13]. Similar concerns are seen on a regular basis as well, it is not uncommon for Radiation Safety Officers at medical facilities to be called on to calculate fetal radiation exposure from diagnostic radiation several times annually, and for many obstetricians to recommend terminating the pregnancy even before a fetal dose calculation has been performed. Accordingly, it seems reasonable to assume that the aftermath of a nuclear attack might lead to similar concerns, resulting in performing unneeded abortions. Much of this is due to the fact that physicians rarely receive significant training in the effects of radiation on their patients or on the reproductive effects of radiation [14].

The last aspect to be discussed here involves evacuations, both those ordered by the government and those undertaken by worried citizens who lack the ability to asses the risks they face on their own. In the aftermath of the 1979 accident at Three Mile Island over 40,000 persons evacuated the area (although the evacuation order should have affected only a few thousand); over 300,000 were required to evacuate following the Chernobyl accident and a similar number evacuated from the areas affected by the Fukushima fallout plume in the weeks after that accident. These evacuations can pose a risk in and of themselves, and the risk increases as the number of evacuees increases.

Another consideration is that, in the event of a nuclear attack, the fallout plume will begin settling to the ground fairly quickly and it will be dangerously radioactive, unlike the plume from a nuclear reactor accident. People trying to flee as their immediate reaction to witnessing an attack do not know where the plume will settle to the ground; in the absence of good information they might well end up fleeing into the plume or along the axis of the plume. Thus, it is possible that those following their instinctive inclination to self-evacuate rather than sheltering might end up *increasing* their radiation exposure.

12.4 Government

Government is never so badly needed as in the aftermath of a terrible event. Emergency responders look to government for direction and to help them to prioritize a multitude of tasks that need to be accomplished; the public looks to government to tell them what has happened and to advise them as to how to stay safe; governmental agencies look to their elected officials to (hopefully) weld them together into an effective force to restore order to the city and to protect the citizens as best they can. "*The public will respond to strong, effective leadership. People need to believe that decisions made are rational*…." [2]. But in the aftermath of a nuclear explosion, government might not be able to act at all, let alone effectively, for some time.

During one meeting this author attended a representative from the Mayor's office indicated the expectation that the government would begin to function coherently in less than an hour after a nuclear attack, including issuing public service announcements instructing the population to seek shelter. In fact, it might take several hours to even have a good understanding of which government officials are even alive, let alone able to convene to begin leading the response efforts or informing the public as to what has taken place. Once government has assembled, it will be necessary for the various agencies—and the various levels—of government to attempt to work in concert with one another rather than competing for resources and responsibilities.

The fact is that a rapid and effective response in the immediate aftermath of a nuclear attack is both essential, and likely to be nearly impossible to carry out. For this reason, it is equally essential that government at all levels develop emergency response plans and to conduct table top and field exercises to practice these plans with the relevant agencies at all levels of government. This process will help to identify the major issues that will need to be addressed, as well as which agencies and which levels of government are best-suited to address them. In addition, by working together on planning and exercises, key personnel will be able to work together under less stressful circumstances rather than during the confusion of an actual response.

Emergency response plans should clearly outline:

- Various **objectives** to be accomplished (e.g. delineate and secure the Hot Zone)
- Major **tasks** required to achieve these objectives (e.g. perform radiation surveys to identify radiation dose rates indicating the Hot Zone boundary)
- Specific **agencies** assigned each of these tasks (e.g. Health and Environmental Protection will establish and adjust radiological boundaries)
- **Constraints** or expectations relevant to accomplishing each task (e.g. Hot Zone boundary will be established at a dose rate of 1 mGy hr^{-1})
- Safety **precautions** (e.g. responders will not exceed a dose of 250 mSv unless engaged in life-saving activities and will not enter any areas where dose rates are in excess of 1 Gy hr^{-1}).

According to Bruddemeier and Dillon [15] the actions that people take in the first several minutes following a nuclear explosion can determine whether they live or die, in a large and densely populated city such as New York City several hundred thousand people will make decisions within the first quarter hour that will help them to survive or that might prove deadly. Considering the likely difficulty of locating and transporting government officials, there might be value in recording some general messages in advance that can be broadcast as quickly as possible following a nuclear attack can help to save the lives of those who are able to receive the message and who have not yet sought shelter.

It will also be important that governmental officials provide a consistent message to the public, while at the same time adjusting their message as appropriate to reflect changing circumstances as more information becomes known (e.g. as radiation levels drop and boundaries contract due to radioactive decay).

As the various levels of government are attending to the responsibilities noted above there will be the temptation to throw all available resources into the area affected. But governments must also consider that the nuclear attack might be the first of a series of WMD attacks across the nation, that it might be followed by additional nuclear attacks, by chemical attacks, or even by simple bombs and shootings. Accordingly, national governments cannot afford to immediately deploy all of their personnel and resources to the city that was attacked; as difficult as it might be, and as much pressure as the public and media might bring, some resources have to be held in reserve in case they are required elsewhere. As time goes by and the likelihood of secondary attacks lessens—and as more is learned about the attack that took place and the group that launched it—government will be able to dig into its reserves of both people and supplies to continue supporting the response efforts.

Government's tasks—at all levels of government—will be seemingly endless in the aftermath of a nuclear attack. It will be important to work to convene the local government as soon as possible to begin directing emergency response efforts and to coordinate its activities with those of the incoming regional and national resources that will be arriving. At the same time, all levels of government will need to communicate with the public (when communications are restored) to ensure that people understand what has happened and how to keep themselves safe. The national and regional governments will also need to provide resources to help stabilize the immediate emergency, while maintaining sufficient reserves in the event of a secondary or follow-up attack elsewhere. Finally, all levels of government will need to be able to coordinate their activities to minimize competition for resources or duplication of efforts.

References

1. Norris R, Kristensen H (July–August 2010) Global nuclear weapons inventories, 1945–2010. Bulletin of the Atomic Scientists. pp 77–83
2. Flynn D, Goans R (2006) Nuclear terrorism: triage and medical management of radiation and combined injury casualties. Surg Clin 86(3):601–636
3. Hall E, Giaccia A (2019) Radiobiology for the radiologist, 8th edn. Lippincott Williams and Wilkins, New York
4. Balicer R, Catlett C, Barnett D et al (2011) Characterizing hospital workers' willingness to respond to a radiological event. PlosONE 6(10)
5. Chaffee M (2009) Willingness of health care personnel to work in a disaster: an integrative review of the literature. Disaster Med Public Health Preparedness 20(1):42–56
6. Karam P (2017) Operation Gotham Shield Exercise. Health Phys News XLV(12):2–3
7. National Council on Radiation Protection and Measurements (2010) Population monitoring and radionuclide decorporation following a radiological or nuclear incident, Report No. 166. NCRP, Bethesda MD
8. Thompson J, Rahn M, Lossius H, Lockey D (2014) Risks to emergency medical responders at terrorist incidents: a narrative review of the medical literature. Crit Care 18:521–530
9. Hall R, Hall C, Chapman M (2006) Medical and psychiatric casualties caused by conventional and radiological (dirty) bombs. Gen Hosp Psychiatry 28:242–248
10. Becker S (2004) Emergency communication and information issues in terrorist events involving radioactive materials. Biosecur Bioterror 2(3):195–207
11. Becker S (2005). Addressing the psychosocial and communications challenges posed by radiological/nuclear terrorism: key developments since NCRP Report No. 138. Health Phys 89(5):521–530
12. Becker S (2011) Risk communication and radiological/nuclear terrorism: a strategic view. Health Phys 101(5):551–558
13. Ketchum L (1987) Lessons of Chernobyl: SNM members try to decontaminate world threatened by fallout. Soc Nucl Med (SNM) Newsline 28(6):933–940
14. Shiralkar S, Rennie A, Snow M, Galland R, Lewis M, Gower-Thomas K (2003) Doctors' knowledge of radiation exposure: questionnaire study. BMJ 327:371–372
15. Bruddemeier B, Dillon M (2009) Key response planning factors for the aftermath of nuclear terrorism (LLNR-TR-410067). Lawrence Livermore National Laboratory
16. World Health Organization (2005) Chernobyl: the true scale of the accident. https://www.who.int/news/item/05-09-2005-chernobyl-the-true-scale-of-the-accident#:~:text=According%20to%20the%20Forum's%20report,danger%20to%20their%20health%20from. Accessed 14 Nov 2020
17. Schwartz H, Flood A, Gougelet R, Rea M, Nicolalde R, Williams B. (2010) A critical assessment of biodosimetry methods for large-scale incidents. Health Physics 98(2)95-108
18. Terkel S. (2007) The Good War: An Oral History of World War II. MJF Books, New York.

Chapter 13
Developing a Radiological and Nuclear Interdiction Network

Unlike biological agents, chemicals, and explosives, radiation can be detected at a distance; a high-activity radioactive source can be detected at a distance of a kilometer or more using hand-held instruments and to distances of more than 2 km with large-volume scintillation detectors. This makes it possible to establish instrumentation networks that can be used to interdict radioactive and nuclear weapons.

As we shall see, an effective interdiction network will include multiple layers; there will be responders with varying levels of training, wielding instruments of varying levels of sophistication and capability, using a variety of techniques, in a number of geographic locations surrounding likely or potential targets.

In most major cities the bulk of the work of interdiction relies on law enforcement personnel walking and driving the streets as they go about their routine duties. One reason for this approach is that most cities have more police officers than radiation safety experts and, if they want to deploy a large number of radiation detectors then it is often most effective to deploy them in the cars and on the belts of the police, whose patrols encompass all parts of the city. However, this means that the majority of interdiction surveys are performed by personnel who are not radiation safety professionals and who are performing these surveys in addition to their routine work. As a result, the radiation detectors given to these personnel tend to be relatively simple and inexpensive devices that the users can be quickly taught to use properly.

This "base" is augmented by lesser numbers of more sophisticated detectors carried by more highly trained responders. And if the city cannot resolve an issue using their own personnel, there is often the ability to contact a national resource (in the US this would be the system of national laboratories and national agencies) for what is called "reachback"—transmitting data to highly trained scientists who are able to contribute their expertise to resolving any problems that arise. In addition, the current trend of using radiation detectors as part of an interconnected network as well as combining information from multiple sensors (e.g. cameras, license plate readers, and so forth) with the information from radiation detectors can provide a great deal of information that can be valuable in interdiction operations.

P. A. Karam, *Radiological and Nuclear Terrorism*,
Advanced Sciences and Technologies for Security Applications,
https://doi.org/10.1007/978-3-030-69162-2_13

The manner in which these all work together and in which one level calls for assistance or "hands off" to a different level of responder is described (in the US) in what is called the Concept of Operations, often shortened to "CONOPS."

There are some questions that must be answered if one is to develop an interdiction network:

- What are the characteristics of the device(s) we are trying to interdict?
- Are we concerned about an internal threat (obtaining a source within the city that will be attacked), an external threat, or both?
- Who will be operating the instruments that will be deployed and what level of training and expertise do they have?
- What is the budget?

Each of these will be discussed in turn.

13.1 Device Characteristics

Alpha radiation cannot penetrate a sheet of paper and has a range of only a few mm in air. Beta radiation can penetrate to a depth of about 1 cm in water and up to six or seven meters in air. Detecting these radiations at a distance can be hard enough when they are unshielded, and sources that emit alpha and beta radiation are even more difficult as they are so easily shielded. Gamma and neutron radiation, on the other hand, are highly penetrating, and gamma radiation has a range of up to several kilometers in air. Thus, gamma- and neutron-emitting sources are detected at a distance, and gamma radiation can be analyzed to identify the radionuclide(s) that emitted it.

A higher-activity source is more difficult to shield and can be detected at a greater distance compared to a source of lower activity. Radiation dose rate drops as the inverse square of the distance to a source and the dose rate at any given distance is proportional to the source activity. Thus, in order to detect Source B at twice the distance as Source A, Source B must have four times the activity.

The device in question might or might not provide shielding for the radioactive source to make it more difficult to detect from a distance. Shielding adds to the size and weight of a device, possibly placing restrictions on the type of vehicle that can be used for transportation and limiting the places where it can be hidden; shielding a high-activity Co-60 source, for example, might require more than a ton of lead. In addition, heavy shielding can absorb the energy of an explosion, reducing the explosion's impact. Thus, radiation shielding can increase the difficulty of detecting a radioactive source, but at the cost of making the source more difficult to transport and reducing the potential effectiveness of the weapon's explosive component.

13.2 Devices Originating Inside Versus Outside of the City

In 2010 this author joined a class of counterterrorism police from New York City in attending a training course on radiological terrorism in Oak Ridge Tennessee; the course was developed for police and security officers and it concentrated on responding to attempts to steal high-activity radioactive sources from various settings.

At what distance can a source be detected?

When developing an interdiction network one consideration is how close together one must place instruments in order to detect a radioactive source of a given activity—the "mesh size" needed for one's interdiction "net." Answering this question is one of the key factors in determining how many detectors will be needed to secure a given event (or a given city) and how much the system will cost to install and maintain. For the purposes of this example, we will assume that background radiation levels are about 0.1 μGy hr^{-1} and that a "detection" occurs when an instrument detects a sustained dose rate of at least two times this level (0.2 μGy hr^{-1}).

Let's assume that our goal is to detect a 500 GBq Ir-192 industrial radiography source. The gamma constant for Ir-192 is 0.0015 mGy hr^{-1} MBq^{-1} at a distance of 30 cm, so a 500 GBq source would produce a dose rate of 750 mGy hr^{-1} at a distance of 30 cm.

Solving the inverse square law to determine the distance at which radiation dose rate from this source is reduced to 0.1 μGy hr^{-1} above normal background levels gives us:

$$R_2 = R_1\sqrt{\frac{I_1}{I_2}}$$

Or, adding in the numbers, $R_2 = 0.3\,\text{m} \times \sqrt{\frac{750\,\text{m Gy hr}^{-1}}{1\times10^{-4}\text{m Gy hr}^{-1}}} = 822\,\text{m}$

Thus, to detect an unshielded 500 GBq Ir-192 source in flat terrain, radiation detectors would have to be placed no more than about 822 meters apart.

In reality, even this spacing might be too great because sufficient thicknesses of air will also absorb some radiation. Thus, an Am-241 source with a lower gamma energy (0.06 MeV) would require a closer spacing between detectors than would, say, an equal activity of Co-60 (gamma energies of 1.17 and 1.33 MeV).

One comment made by the first instructor was particularly noteworthy; "A terrorist who steals a source in New York City doesn't have to avoid detection for a few days while they transport the source to their target; they only need to get it outside because *they're already at their target*." The point was that a radiological attack might originate from within the target city, making interdiction both more difficult and more important due to the truncated timeline. Thus, when developing an interdiction network one must consider not only the threat of an RDD or nuclear weapon brought into the city from outside, but also the possibility that a source might be obtained from a location within the city limits. Having said that, it seems safe to assume that a nuclear device would originate outside the city since these weapons are not normally stored in urban areas.

13.2.1 Interdiction for Sources Obtained Within the City

Interdicting radioactive sources that are stolen from within the city can be difficult; it requires securing the sources against theft as well as possible and installing alarm systems to alert Security and law enforcement forces so that they can react quickly to prevent the source from leaving the building in which it is stored. This can include installing cameras, intrusion detection devices, radiation detectors, and the like, transmitting alarms to one or more central locations (e.g. Security office, the local police station, etc.).

Having security systems is a good start, but they are insufficient unless facility Security forces and/or the local law enforcement personnel can respond to alarms quickly and appropriately. This was the reason for the training provided at Oak Ridge, to give these two groups the opportunity to not only respond to alarms in an exercise setting, but to also have the opportunity to train together to better coordinate their response.

It is also important for those who own high-activity sources to examine their existing source security systems and procedures to ensure their sources are adequately protected; if sources are adequately protected against theft then a response will not be necessary. The New York City Department of Health and Mental Hygiene helped to develop a series of tools—checklists and self-audit forms, for example—to assist organizations with high-activity radioactive sources to determine if they are adequately protected and actions that can be taken to help better secure them [7]. More recently, the US Nuclear Regulatory Commission developed similar guidance for radioactive materials licensees [10].

13.2.2 Interdiction for Sources Obtained Outside the City

Trying to interdict radioactive or nuclear materials being transported into a city from elsewhere poses challenges as well [3]. Consider, for example, a scenario in which a radioactive source is reported as having been stolen from a facility in Harrisburg, Pennsylvania. Major cities close to Harrisburg include Washington DC, New York City, Baltimore, and Philadelphia. Between Harrisburg and these four cities is about 1000 km of highway and several thousand km of secondary roads; the area through which the source might be transported or hidden is over 50,000 km^2.

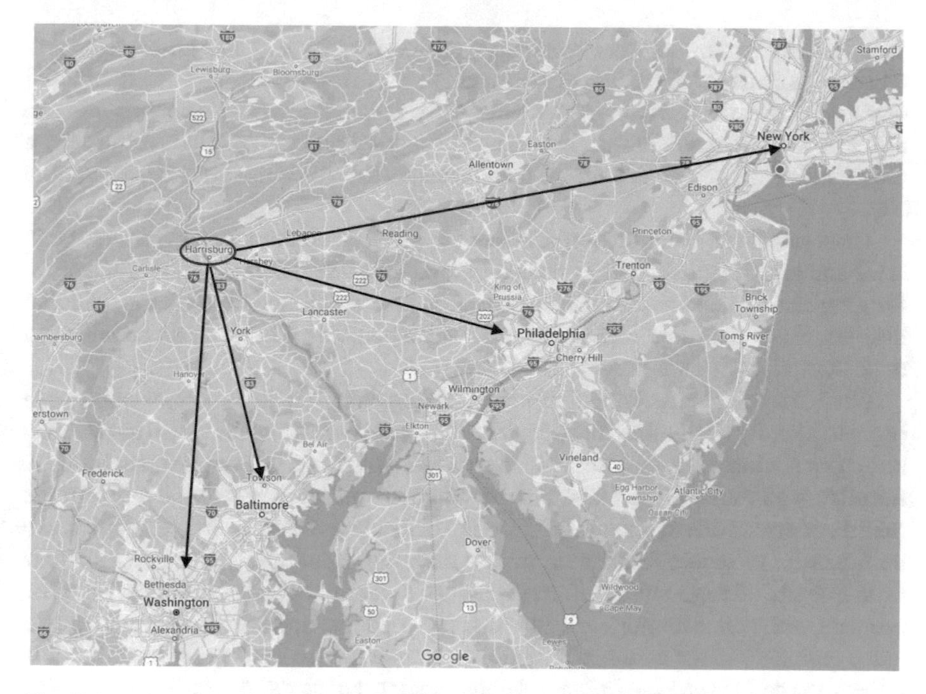

Fig. 13.1 A map of the area discussed in the text. The many roads in yellow are the highways; there is also a network of secondary roads (white) as well as thousands of additional roads not shown here

Imagine being the person tasked with coordinating efforts to find this source before it can be used in an RDD. With hundreds of kilometers of highway and thousands of kilometers of secondary roads to between Harrisburg and potential targets, how best to use limited resources? (Fig. 13.1)

Available resources are the existing radiation detectors and the personnel operating and monitoring them, and the detectors are likely a combination of fixed and portable instruments:

- Fixed monitors at choke points and check points (e.g. service areas, weigh stations, toll plazas, and so forth)
- Vehicle-borne systems to cover the main roads
- Aerial systems to survey large areas quickly
- Hand-held instruments used by personnel deployed to high-priority areas and/or responding to alarms on other systems
- Radioisotopic identification instruments to help adjudicate alarms.

The person(s) coordinating interdiction activities will likely deploy these instruments in a manner so as to take advantage of their respective strengths and weaknesses.

For example, aerial instruments are highly sensitive and allow for nuclide identification, but they cannot help identify the precise location of a radioactive source;

handheld PRDs, on the other hand, are very good at identifying the precise location of a radioactive source but are not nearly as sensitive as the large-volume aerial detectors and cannot survey large areas nearly as quickly. Aerial detectors are also able to surmount many obstacles that would slow or prevent an effective survey on the ground—walls, ravines, un-bridged rivers, confusing mazes of streets, and the like.

Thus, an effective interdiction plan will use aerial detectors to quickly survey large areas and long lengths of highway, conveying the location of any "hits" to teams on the ground who can localize the source using their hand-held detectors. Additional instruments (waterborne, vehicular, drone-mounted, and so forth) can be added to the interdiction effort as they become available.

While the initial interdiction efforts are taking place, the interdiction team(s) can begin to identify (if this has not already been done) choke points at which to establish screening stations. These would include bridges and tunnels through which traffic is funneled—all traffic into Manhattan, for example, must come through one of three tunnels or across one of sixteen bridges; if Manhattan is thought to be a target, it can be far easier to secure fewer than twenty crossings than to try to secure hundreds or thousands of large and small roads across thousands of square kilometers. The problem with concentrating on these final chokepoints, however, is that they are all at the periphery of the assumed target; these should be the last in a *series* of barriers rather than comprising the primary defense against an attack. Accordingly, it makes sense for interdiction efforts to include chokepoints that are further away from the presumed targets; toll road entrance and exit ramps, toll plazas (although the advent of high-speed cashless tolling stations reduces the efficacy of these locations), service areas (especially on limited-access roads), major bridges, and the like.

In addition to taking measures to interdict the stolen materials while they are in transit to the target, the potential target cities can begin to establish their own interdiction networks. These will likely involve the use of the same types of instruments that are being deployed for the large-scale interdiction efforts—at least, up to the capabilities of the various cities.

Establishing an interdiction network within a city, to protect critical infrastructure, an iconic location, or a special event, is similar to the process described above, although there are some differences. The primary difference, of course, is that there is no time pressure, with the corollary being that the network might have to operate for extended periods of time—for days, weeks, or even years. New York City, for example, has been managing radiological interdiction networks for major events (e.g. Time Square at New Year's Eve, the United Nations General Assembly, and many others) every year since 2003 and have been conducting radiological interdiction surveys on a routine basis since before 2010.

Another difference between monitoring a city for radioactive sources originating within the city is that the target city is known; it is not likely that a terrorist group would steal a source from a hospital in New York City to use in an attack against Philadelphia. Any one city has a limited number of high-activity sources that tend to be in fixed locations; such sources are easier to secure against theft, and those who own these sources are able to work with law enforcement agencies to determine in advance how various incidents will be addressed. In addition, local law enforcement agencies can conduct exercises with facility security personnel to practice working together, and can carefully select locations best suited for fixed monitors, mobile systems, observation posts (if appropriate), and other monitoring and security capabilities.

In the case of Singapore (as well as many other major cities) there are only a limited number of access points; any high-activity source coming into the city will have to come across one of two bridges or by boat, and the size and weight of the shielding will limit the number of places along the shore to those close to highways. Thus, monitoring strategic intersections and choke points as shown in Fig. 13.2 (especially near the city's downtown core) as well as the bridges and docks will additional assurance. Supplementing these fixed locations with large-volume mobile detectors in helicopters, boats, and vehicles in addition to cops carrying PRDs will provide a robust interdiction network. This can also be done on a smaller scale, as shown in Fig. 13.3.

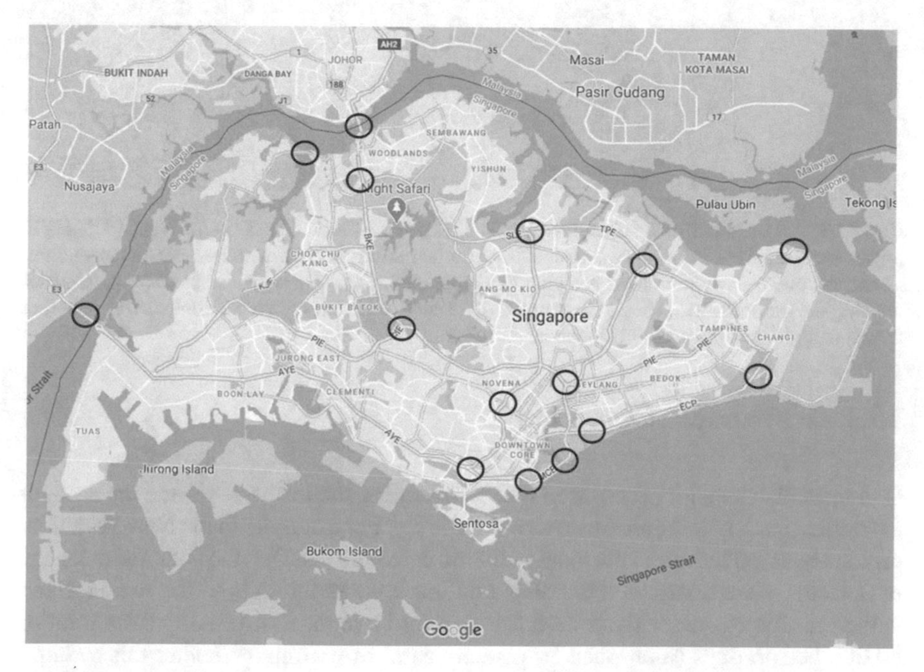

Fig. 13.2 An example of a city-wide interdiction network. The stations noted here would be augmented by mobile systems (aerial, maritime, and vehicles) and cops with PRDs

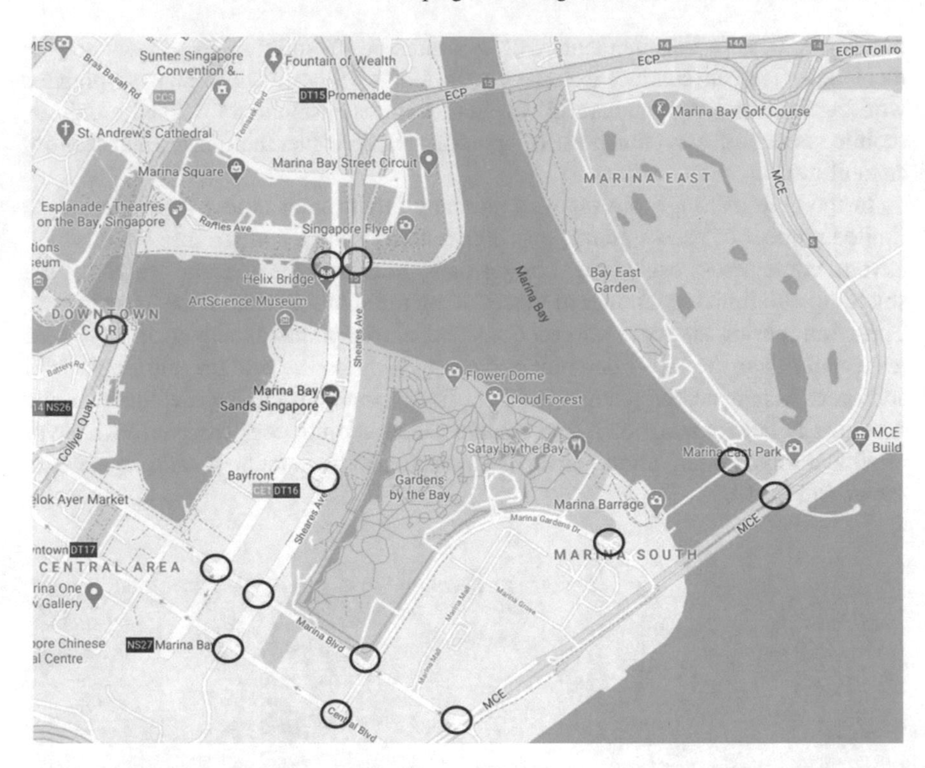

Fig. 13.3 An example of a single-location interdiction network. For a single event (e.g. a VIP appearance) the fixed locations would consist of portal monitors, large-volume detectors in vehicles at choke points, and individuals carrying backpack detectors. There would be cops with PRDs walking the grounds, with special attention to the waterfront, along with helicopter surveys before the event and maritime patrols throughout

13.3 Instrument Operators

One must also consider those who will be operating the instruments and the amount of training each will require to do so competently. For example, in the New York City region about ten thousand personal radiation detectors (PRDs), the simplest category of detector, were distributed to police and fire fighters throughout the region. Each person receiving one of these detectors received two days of training for a total of 20,000 person-days (more than 50 person-years) of training. Whether this training is received during normal working hours or as overtime (many American police departments perform training on overtime, paying workers a 50% or greater premium on their normal wages), it represents a cost that must be factored into the cost of the instruments themselves.

One system in use in the US (e.g. [4]) recognizes that more sophisticated instruments require more training in order to use them and to interpret the results properly. A Tier 3 responder, for example, is one who receives the most basic training and who is given the most basic instruments. A Tier 3 emergency responder will be given a simple PRD that measures only radiation dose rate, and they will be trained on how to distinguish between, say, a car carrying a nuclear medicine patient versus a car carrying a radiological device. To adjudicate such an alarm they will rely on what police term "the totality of circumstances," using a combination of their instrument readings, the person's age and demeanor, the location where the stop was made, response to questions (including consistency of answers and the agreement of these answers with other observations), and so forth.

If the totality of circumstances indicate that the person is a harmless nuclear medicine patient then the alarm would be considered adjudicated; if not then the officer might call for a Tier 2 responder to help collect additional information. Consider for example, a person claiming to be a nuclear medicine patient who is carrying a backpack; when separated from the backpack by a few meters, radiation dose rates near the "patient" drop to background while the backpack remains "hot." This would suggest that the radioactivity is contained within the backpack, not within the "patient" as one would expect. This would call for follow-up.

A Tier 2 responder might be equipped with a radioisotope identifier (RIID) to identify the radionuclide(s) present. If, for example, the nuclide in the backpack was found to be I-131 then it is possible that the person really is a nuclear medicine patient and that the backpack was contaminated at the hospital. If the backpack contained Ra-226 then it might contain an RDD…or it might contain the person's radioactive rock or antique watch collection. If the backpack contains Cs-137 it might be a "button" source for checking radiation instruments or it might be that the backpack was contaminated as the person was building an RDD. Identifying the nuclide(s) present, in conjunction with additional questions from a better-trained officer, can help to determine if the person is free to go, or if they should be questioned and investigated further.

There are times that even a Tier 2 responder will not be able to resolve a situation; in these cases it might be necessary to call for a Tier 1 asset. In the US the Tier 1 assets are the National Laboratories, who will carefully analyze the spectra sent to them through a process called "reachback." In the example above, a Cs-137 identification would not call for reachback as the Cs-137 spectrum is unambiguous. On the other hand, the Ra-226 gamma has an energy almost identical to that of U-235 (186 versus 185 keV) and it is not uncommon for a RIID to mistake one nuclide for the other. Thus, if Ra-226 was identified, it would be prudent to send the spectrum to a Tier 1 asset for assessment, to confirm the nuclide identification and the material giving rise to it.

Quantifying interdicted sources

During a three-state radiological interdiction exercise conducted in a large American city in 2011 the scenario called for the "theft" of a high-activity (about 90 TBq) Co-60 radioactive source from a facility in a neighboring state. Over the course of the next two days, investigators determined that the source had been divided into multiple smaller sources that were being sent to attack a variety of targets. As they "caught" the vehicles transporting these sources the question that quickly arose was "Have we accounted for all of the stolen radioactivity or is there more out there?"

The first "hit" called in was from a helicopter, and the instrument operator reporting a dose rate of 0.27 mGy hr^{-1} while hovering directly above the vehicle with the source at an altitude of 50 m.

Using the inverse-square law we can easily determine that the dose rate at a distance of 1 m will be $50 \times 50 = 2500$ times as high as the dose rate measured by the helicopter; $0.27 \times 2500 = 675$ mGy hr^{-1}.

The gamma constant for Co-60 is 8.53×10^{-17} Gy s^{-1} Bq^{-1} at a distance of 1 m. Dividing the calculated dose rate at 1 m by the gamma constant and doing appropriate unit conversions (e.g. 3600 s in one hour, 1000 mGy in one Gy, and 10^{12} Bq per TBq) shows us that the detected source has an activity of 2.2 TBq, assuming the source is unshielded. Thus, while this is a substantial source, it represents only a fraction of the missing radioactivity.

As the search continues, the following sources are identified: (Table 13.1)

Thus, assuming that none of these sources are shielded, 31 TBq remain to be found.

An emergency response organization with, for example, 1000 personnel might distribute PRDs to 100 Tier 3 responders, while maintaining five to so RIIDs to be used by any of ten or so Tier 2 responders, each of whom has received several days of training on their equipment. Some cities will also maintain additional detectors if they are especially interested in (or concerned about) radiological security, if they are located near a large radiological or nuclear facility, if they host occasional high-profile events; these can include detection systems (vehicular, aerial, or maritime), backpack detectors, portal monitors, or detectors in fixed locations.

It is worth mentioning that, while the majority of the instruments given to Tier 3 personnel have been designed to be easily used and interpreted by emergency

Table 13.1 Adding up the activity of the sources found

Hit #	Dose rate (mGy hr^{-1})	Distance (m)	Activity (TBq)
1	0.27	50	2.2
2	0.90	20	1.2
3	22	5	1.8
4	32	2	0.42
5	1	45	6.6
6	47	15	34
7	6.0	25	12
			59

responders, many of the more sophisticated instruments used by Tier 2 responders are not; it is not unusual for a mobile system or a hand-held RIID to be designed for scientists or for radiation safety professionals, to require extensive practice in order to operate properly and maintain proficiency, and to display information in a manner that is not intuitive. Emergency responders can learn to operate these instruments and to use them well, but their supervisors and managers must understand that achieving and maintaining these skills requires both initial training and continuing practice lest these skills be forgotten.

13.4 Instrument Selection

Here's a story for you. In the 2010 s a senior police officer in a major American city had a brainstorm: "We should put radiation detectors on buildings!" He took this idea to his management, explained his idea to them, and the police department secured funding to implement the idea. They contacted a company with whom they had done a great deal of business in the past and the company's representative replied "We have just the thing for you!" After spending time on the requisite negotiations and contracts, the company began installing their radiation detectors on buildings in the city. After several years over 30 units had been installed and their data lines tied into the city's central network (which included databases, video cameras, and a number of other sensors) and the city was beginning to negotiate the purchase of an additional 70 detectors.

At this point, a radiation instrumentation expert was hired by the city and, among other tasks, was asked to evaluate the radiation sensor network. After several months the radiation professional asked what purpose the sensors were to serve:

- Interdiction
- Health and safety monitoring during a radiological emergency
- Tracking fallout plumes
- Providing information for dose reconstruction and long-term public health concerns.

He then pointed out that the detectors were of a type (gamma scintillators) that were suitable for interdiction, but their location (at the tops of buildings) was not. They overloaded at dose rates of a fraction of 1 mGy hr^{-1}, making them far too low-range to be useful for health and safety or for tracking fallout plumes. And they were too inaccurate—especially when measuring any nuclides other than Cs-137—to be useful in reconstructing radiation exposures from, for example, mixed fission products or low-energy gamma-emitters. Thus, while each unit was an excellent detector in and of itself, the network as a whole was virtually useless because the customer had never discussed with the vendor what the network was to be used for. Had they had this discussion, the detectors that were finally installed would have been far different. In fact, subsequent conversations with the vendor led to the testing of a different system that include two energy-compensated GM tubes that made it possible to perform all of the tasks listed above except for interdiction (due to the location of the detectors), and changing their location from rooftop to street level accomplished that. By this time, however, the grant under which the detectors were being purchased had expired, leaving the city with a small network of detectors that were unable to serve any useful function, and without any of the more capable detectors.

The moral of this particular story is that, before selecting any radiation detectors, one must know what it is that they are to be used for and to communicate this clearly to the vendor(s) that manufacture and sell them. This is why it can be helpful to consider what the city plans to do with the radiation detectors—the mission they are trying to accomplish—as well as who will be operating them when trying to decide which detectors to purchase (e.g. [4]).

Consider the potential uses noted above and the characteristics of a detector to meet the particular needs of that city.

- **Interdiction** requires a high degree of sensitivity to detect even low-activity or well-shielded radioactive sources. The ability to identify radionuclides is also very useful. Detectors used for interdiction should be placed as close as possible to the locations they are intended to survey to minimize attenuation due to distance. Sodium iodide, cesium iodide, and other gamma scintillators are idea for this purpose due to their relatively low cost, relative ease of operation and upkeep, and the lack of need for liquid nitrogen.
- **Health and safety monitoring** requires the ability to accurately measure very high radiation dose rates, in excess of 1 Gy hr^{-1} so that emergency responders and those supervising them are aware when they enter areas with dangerously high radiation levels from one or more radionuclides. Thus, a detector for such a purpose should be energy-independent. This purpose is best served by an ionization chamber or energy-compensated GM detector. These devices should be placed at ground level, where emergency responders will be working and through which members of the public will be evacuating.
- **Tracking fallout plumes** also requires the ability accurately measure both relatively low and very high radiation dose rates from nuclides with a variety of gamma energies. It does not require a high degree of accuracy, although accurate readings are always better than readings that are inaccurate. Here, too, ion chambers and/or energy-compensated GM detectors are ideal, although it might be necessary to use multiple detectors to provide reasonably accurate measurements from dose rates that vary over several orders of magnitude, from a few tens of μGy hr^{-1} through a few tens of Gy hr^{-1}. Tracking and mapping the footprint of a fallout plume is best accomplished using detectors on the tops of buildings to minimize the effects of shielding by high-rise buildings as well as to avoid local effects caused by winds swirling around buildings and other structures at ground level.
- **Dose reconstruction and the later adjudication of radiation injury compensation claims** calls for the ability to measure accumulated dose with a high degree of accuracy, ideally from multiple energies of gamma radiation and (in the event

of a nuclear attack) neutron radiation, at dose rates ranging from near-background to several Gy or Sv hr^{-1}. This can be accomplished with a combination of pressurized and air ionization chambers, with a series of energy-compensated GM detectors, or with any of a number of radiation dosimeters (e.g. direct ion storage or TLD); neutron dose can be measured using suitable neutron detectors or neutron dosimeters. Detectors to be used for dose reconstruction are best placed as close as possible to the location of those whose dose is to be determined; in most cities this would be at or near ground level, but cities in which a large number of people live and/or work in skyscrapers should consider placing some detectors or dosimeters in high-rise buildings as well, perhaps 100-200 meters above ground level.

13.5 Budgetary Considerations

The financial impact of a radiological attack is roughly proportional to the amount of radioactivity used [1, 2]; the size of the contaminated area, the radiation dose rate and subsequent health risk, the length of time required for the radioactivity to decay to acceptable levels...all of these are proportional to the source activity. Thus, the cost of an attack will—very roughly—be proportional to the amount of radioactivity present, with high-activity sources being generally more costly than low-activity sources of the same nuclide.

Zimmerman and Loeb [12] estimated the cost of a radiological attack against a major city to be on the order of tens of billions of dollars for source containing several TBq of radioactivity. If we assume that the total cost of an attack is roughly proportional to the amount of radioactivity used then this implies that an attack using several GBq would inflict a cost in the tens of millions of dollars and an attack using several MBq would cost tens of thousands of dollars to recover from. While the psychological cost of an attack cannot be easily quantified, it seems reasonable to feel that it might not make sense to spend, say, hundreds of millions of dollars to establish an interdiction network capable of detecting sources that would lead to only tens of thousands of dollars in damage and other costs.

The distance at which a given detector can "see" a radioactive source is proportional to the square root of the source activity. Thus, if an interdiction network requires, say, 100 detectors to stop the movement of a 10 GBq source, it will require 10,000 detectors to stop the movement of a 1 GBq source. In other words, attacks that cause the least damage will be the most expensive to stop. Each city must determine how much it can afford to spend and what level of attack (or malicious misuse) it is willing to accept.

In addition to the initial cost of purchasing radiation detection equipment the department and city must also consider the costs of training the Tier 2 and Tier 3 users (both initial and periodic refresher training), instrument maintenance and calibration, and the replacement of instruments that break or become obsolete. As this book is being written the National Council on Radiation Protection and Measurements (NCRP) is developing guidance on calibration and other instrument maintenance that should be released shortly after this book is published. These costs are conservatively estimated to be at least 10% of the initial purchase cost each year (assuming an average instrument lifespan of about 10 years), and likely more, so an initial investment of, say, $5 million will require an annual expenditure of about $500,000 to pay for maintenance, calibration, replacement, training, and other incidental costs [1]. As a corollary, without budgeting for these regular—and continuing—expenditures, even a robust radiological detection capability will dwindle to the point of being ineffectual over a period of a decade or less. Interestingly, some large police departments have elected to establish their own in-house calibration capabilities to help defray these costs; the New York City Police Department, for example, spent about $250,000 to set up a calibration laboratory, saving about $200,000–$300,000 in annual calibration and repair costs [5].

13.6 Other Factors to Consider During an Interdiction/Investigation

Consider a police officer walking or driving along a street when their radiation detector alarms. They are able to determine that the radiation is coming from a dumpster or trash bin and call for a Tier 2 responder with a RIID; upon arrival the radionuclide is identified as Cs-137. What does this mean, and how should the police respond to this discovery?

There is only one completely innocent explanation that readily comes to mind—that a professional radiation worker disposed of an exempt Cs-137 source (perhaps an instrument check source) into the trash, although this odds of this being the correct explanation are small as so low-activity a source would be hard to detect at more than a meter or so distance. Other explanations include:

- Illegal disposal of radioactive materials
- Theft and disposal of a soil density or other relatively low-activity source
- Theft and disposal of an industrial radiography or other relatively high-activity source
- Worker who became contaminated in an accident
- Contamination from the construction of an RDD.

Most of the possibilities require a degree of follow-up investigation by either regulatory or law enforcement agencies.

There is also the possibility that personnel working with radioactive sources might develop radiation injuries or radiation sickness and might report to a local hospital to seek medical care as was observed following the inadvertent theft of a high-activity Co-60 source in Mexico in December 2013 [8]. Thus, if law enforcement and/or counterterrorism agencies have intelligence of a possible pending radiological attack they should consider asking local physicians, hospitals, and medical clinics to report patients with suspicious injuries or illnesses that could be due to radiation exposure. (Note: nuclear weapons materials are only weakly radioactive and are unlikely to cause radiation burns or radiation sickness).

Initial entry into a radiological area

Ideally, the initial entry into a radiological area should be made by radiation safety professionals; if there is the possibility that explosives or other hazardous materials might be present then the room should also be cleared by a bomb technician, hazardous materials technician, or another worker with the relevant skills.

If the person in charge feels that the situation requires an immediate entry, one technique that is taught is to enter the room and survey in a "Z" pattern, making five quick radiation measurements (one in each corner and one in the center). The person performing the survey should be dressed in appropriate PPE (Personal Protective Equipment) and should avoid touching or handling anything seen in the room. At each survey location the person conducting the survey should wait about 30 s for their meter readings to stabilize and then tell the person recording information (by voice, phone, or radio) the survey location and instrument reading as well as the location and nature of any other possible hazards or items of interest. As soon as the final data point is surveyed the surveyor should exit the room as quickly as possible. Upon exit, personnel present should review the survey map and data, using that information to plan their next entry (Fig. 13.4).

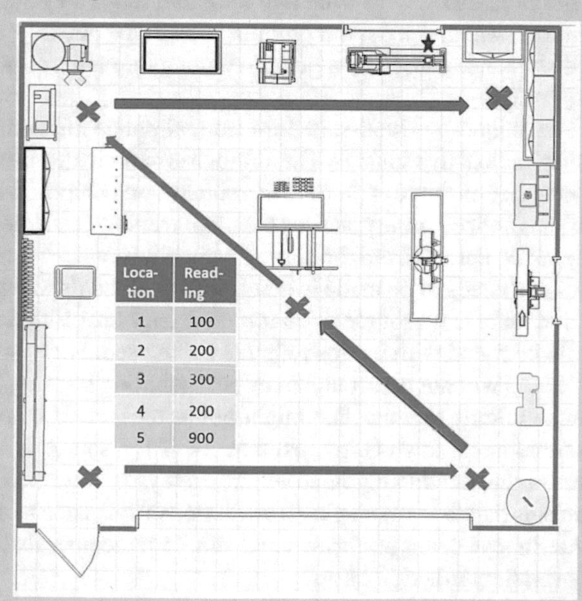

Location	Reading
1	100
2	200
3	300
4	200
5	900

Fig. 13.4 Example of the "Z" pattern initial scoping survey

13.7 Conducting Interdiction Surveys

Among the first things a radiation worker is taught is how to perform an effective radiation survey. They are taught this because we cannot sense radiation or radioactive contamination, their instruments are their "eyes" and if they do not use their instruments properly they will be blind to even high levels of radiation or contamination. They are taught, too, the importance of performing disciplined surveys, paying close attention to the speed of the detector, its distance from the area (or object) being surveyed, and they are trained to pay close attention to covering as much of the area being surveyed as possible. While there are times requiring a quick survey, these occasions are rare and, even then, the initial scoping survey is almost invariably followed by a more careful and thorough survey (e.g. [9].)

Police work is quite different. While police will perform painstakingly thorough investigations of a crime scene, this only constitutes a small part of routine police work. A police officer searching for a missing child, for example, needs to quickly assess an area to see if the child (or a suspicious-looking adult) is there and, if not, to quickly move on to the next location. Pilots who are searching for a missing elder realize that a person on foot does not move very quickly so they tend to circle the area where the person was last seen, using their eyes and their cameras to comb the area for the face (or the clothing) they've been told to search for. If they are searching for an escaping criminal, whether from the ground or from the air, they might well be looking for the disturbance caused by a person running through a crowded sidewalk or darting across traffic.

These search techniques are well-suited for normal police missions, but they are not ideal for finding radioactive sources. The author has participated in a large number of radiological interdiction surveys on foot, in vehicles, on boats, and in helicopters. In each setting he has noticed some practices that, while understandable in the context of normal police work, detract from the survey's effectiveness.

Another point that needs to be made is that a survey that fails to turn up a source does not mean that there is no source to be found, it only means that there were no sources that could be found with the instruments being used. For example, a search with gamma detectors will not detect alpha- or beta-emitting sources. Not only that, but even a gamma-emitting radionuclide might not be detected if it is a low-activity source, if the source emits low-energy gammas, or if the source is well-shielded. This does not mean that conducting interdiction surveys is worthless or a waste of time, it simply means that those managing interdiction efforts need to understand the limitations of the surveys being performed and should remember that they are only one part of the overall interdiction effort.

13.7.1 Survey Speed

When surveying for radiation the general rule of thumb is that slower is better; a radioactive source must be within range of the detector long enough to be "seen." If a detector moves too quickly over a source it might only detect a handful of counts that might simply look like a brief fluctuation in background count or dose rates. In addition, the further one is from a radioactive source the more slowly one must survey because radiation levels drop with the inverse square law. For this reason, aerial surveys using high-volume sodium iodide detectors are typically done at an altitude of about 100 m and a ground speed of 60–100 knots (about 100–200 km hr^{-1}). If a possible source is detected from a helicopter the pilot can slow down or even hover to try to identify the radionuclide(s) present and to narrow down the possible location; fixed-wing aircraft, unfortunately, cannot hover and can only slow down to about 100–150 km hr^{-1} without falling out of the sky. This can place limits on the lower activity of a source that can be detected by a fixed-wing aircraft.

Performing a survey from a vehicle also requires a relatively modest survey speed; at the same time, driving too slowly can slow down traffic and can even be dangerous, so speed limits and traffic patterns can interfere with survey quality—performing a survey from a vehicle on the highway can be even more difficult. Surveying on foot is unlikely to proceed so quickly as to miss a radioactive source. However, it is not uncommon for personnel who are doing a walk-over survey to slow considerably when they first see elevated radiation dose rates or receive an alarm. Unfortunately, given the large distances at which a moderate- or high-activity source can be detected, this can lead to an excessive delay in reaching the location of the actual source. Consider, for example, a police officer who slows to a very slow speed once their radiation detector alarms…on a source that is over 1 km distant…compared to a police officer who continues walking at a normal pace as they watch radiation dose rates increase and then drop again when they pass the source. In the first case it might take hours to reach the source while, in the second, the approximate location will be found in less than a half hour.

13.7.2 Survey Pattern

In the majority of cases the most certain way to find a source is to conduct a survey according to a plan, and that plan will typically involve surveying according to a pattern. Following that survey plan—and survey pattern—will usually take time, but is certain to ultimately find any detectable source in the search area.

The easiest search pattern to follow is a simple grid search—traveling parallel search lines spaced such that the lowest-activity source of concern can be detected from halfway between the search lines. So, for example, the 500 GBq Ir-192 source noted earlier in this chapter, with a detection distance of 882 meters, would call for search lines as much as 1750 meters apart. In reality, however, it is reasonable to

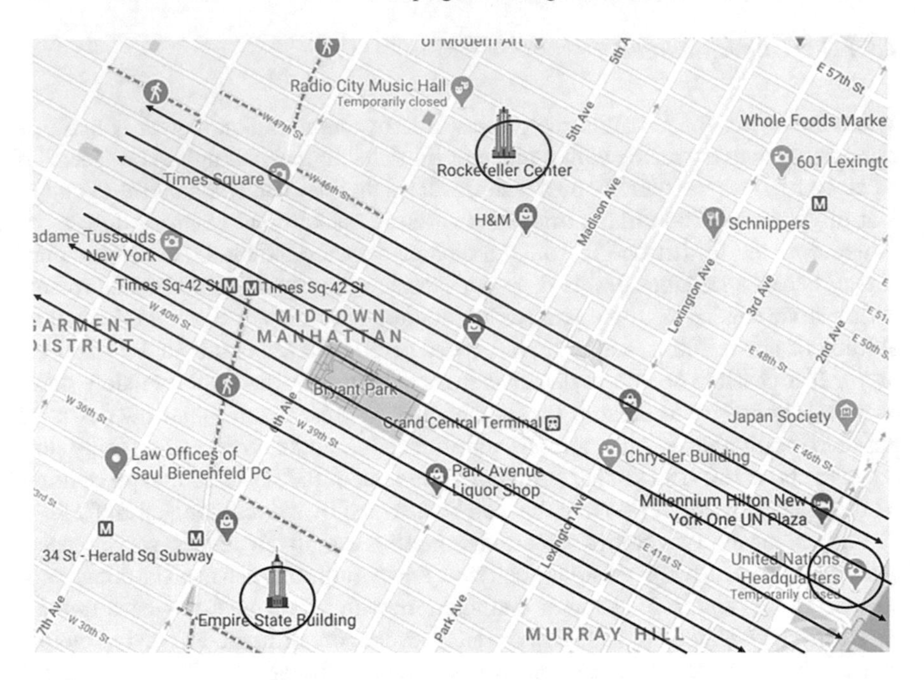

Fig. 13.5 Example search pattern for Midtown Manhattan, including both a grid search from 38th to 45th Streets between the East River and 8th Avenue with additional searches in the vicinity of the Empire State Building, the United Nations complex, and Rockefeller Center

assume that a source is in a shield and that additional shielding is present in the form of the structure in which the source is located and any other structures that lie between the source and the helicopter or vehicle conducting the search. Thus, a more realistic search pattern would simply call for cars to drive along each street or for helicopters to fly above the middle of each block. This not only helps to address the effects of shielding, but also increases the chance of identifying any radionuclide(s) found (Fig. 13.5).

Other search patterns may be used, depending on the circumstances: an expanding square pattern can be useful when the approximate location of a source is known, a sector search pattern can be used to search locations in which only a few areas might be of concern, and other patterns can be used as appropriate. What is important is that a search pattern is used to reduce the chance that a source might be missed.

One possible exception to this would be a targeted search in a limited area. This could be the result of intelligence or an investigation that has identified a specific site—a storage area for example—or it could be that the region being searched has only a limited number of targets that make sense to search. An example of the latter could include a harbor or anchorage, or even ships that are overflown as they approach port. An individual container ship might be huge if searched on foot, it can be large if the survey is done from a harbor patrol boat cruising alongside the ship, and appears fairly compact when surveyed from the air by an airplane or helicopter. Thus, a container ship might call for no survey pattern when surveyed from the air, a careful survey of both port and starboard sides if surveyed from a boat, or a detailed survey plan if surveyed by a boarding party. Containers pose additional challenges when searching for nuclear weapons owing to the low radiation emissions from fissile materials [11]; these will be discussed in the following section.

13.7.3 Survey Discipline

Whatever survey pattern is used, those conducting the survey should have the discipline needed to stick with it unless they have a very compelling reason to deviate from the plan (e.g. visually sighting a box with radiation markings on it in the distance in the direction of the highest readings).

Consider a police officer conducting a survey who, upon receiving an alarm, immediately stops, attempts to identify the direction of the source by turning in a slow circle, and then walks slowly in that direction.

- A high-activity source might be more than a kilometer away; if the survey proceeds at a slow pace, it might require more than an hour to reach the source's location.
- Radiation from a high-activity source can be scattered in an urban environment filled with buildings of various construction, parks, squares, streets, alleys, and so forth. The reading at any given location might not be representative of the radiation environment at that distance from the source and can be difficult to interpret properly.
- Repeated parallel traverse lines might be required to get an accurate idea of a source's location.

Thus, the best option is to continue performing a disciplined survey—if possible, augmented by additional ground, maritime, or aerial assets—until enough information has been collected to determine the source's location.

If there is a degree of urgency about finding the source—if, for example, intelligence sources indicate that an attack is imminent—it is better to put additional assets into the field with all of them conducting well-planned and disciplined surveys as parts of a coordinated interdiction effort overseen and monitored from a central location than to have each survey team operating more or less independently in the manner they feel best.

13.7.4 Monitoring Surveys

Finally, it is unrealistic to assume that any single surveyor or survey team will be able to place their own efforts into the larger "big picture" of the entire survey. Each person or team knows only what they can seem or what they hear over the radio. For any large-scale effort there is no substitute for having a single location—a single display—at which all survey information is consolidated and monitored by a single person knowledgeable enough to make sense of the information.

For decades, surveying large areas—especially in urban environments—was difficult, time-consuming, and interpretation of the results might require still more time. Most readings were taken by individuals looking at their instruments and recording them on survey maps or in a table; this hand-recorded information would be brought to a central location and assembled into a map. Even information recorded electronically was not consolidated until all of the individual electronic records could be brought together and plotted.

In recent decades, and especially in the second decade of this century, this began to change. The advent of ubiquitous GPS, "smart" phones and tablet computers, wireless communication, and ever-more sophisticated software has made it possible to collect data from a large number of different types of instruments, consolidate it at a single location, and to display the entire data set to anyone who can log into that central location. Not only that, but instruments that were until recently the province of highly trained specialists (e.g. gamma spectroscopy) are now designed to be used by non-specialists with little training and to give the ability to forward problematic information to specialists who might be working hundreds or thousands of kilometers away.

Further, some software systems (e.g. RadResponder, maintained by the American Federal Emergency Management Agency) will accept information from a wide variety of instruments and will display it to anybody with permission to view a particular data set. Thus, data collected near Times Square by a New York City Health Department scientist can be viewed by a Chief with the New York City Police Department at the NYPD Emergency Operations Center at One Police Plaza in lower Manhattan, by the Fire Department's on-scene Health and Safety Officer, and by Federal officials in Washington DC who are readying their personnel and equipment to send to New York to assist with response efforts. This ability to collect, analyze, and disseminate information on a nearly real-time basis is likely to increase with time and will enhance the ability to monitor, coordinate, and respond to radiological alerts, incidents, and emergencies as they arise.

While everybody involved in interdiction is looking for illicit radioactive materials intended for malicious use, it is important to remember that interdiction operations have consistently resulted in a number of "innocent" alarms from:

- Nuclear medicine patients
- Buildings and other structures with granite facades
- False alarms from momentary fluctuations in natural background radiation
- Medical facilities and/or radiopharmaceutical delivery vehicles
- Areas fertilized with high-potassium or super-phosphate fertilizers.

Cities that have taken the time to perform comprehensive background radiation surveys will identify some locations and/or facilities that tend to set off radiation alarms; cities that have not had the opportunity to do so will have to adjudicate these alarms as they occur. It is important that those performing interdiction surveys not be distracted or fooled by innocent alarms, and that they be able to distinguish between innocent alarms and events that require investigation.

References

1. Almemar Z, Karam P (2017a) Considering an actuarial approach to prioritizing radiological counterterrorism expenditures. Fed Am Scientists Public Interest Rep 69(5):28–32
2. Almemar Z, Karam P (June 2017b) Measure for measure. CBRNE World 50–53., Falcon Communication Ltd
3. Cazalas E (2018) Defending cities against nuclear terrorism: analysis of a radiation detector network for ground based traffic. Homeland Security Affairs 14(10)
4. Federal Emergency Management Agency (2017) Resource typing definition for screening, search, and detection prevention: preventive radiological nuclear detection team. https://rtlt.preptoolkit.fema.gov/Public/Resource/ViewFile/0-508-1178?type=Pdf&s=Femald. Accessed 10 December 2020
5. Karam P (2016) Setting up and Operating the NYPD radiation instrument calibration facility. PEP W-4, Health Physics Society Annual Meeting, Spokane WA, July 17–21
7. New York City Department of Health and Mental Hygiene (2013) Best practices for securing radioactive materials
8. Romo R, Parker N, Castillo M (4 Dec 2013) Mexico, Stolen Radioactive Material Found. CNN https://www.cnn.com/2013/12/04/world/americas/mexico-radioactive-theft/index.html Accessed 17 Nov 2020
9. US Department of Homeland Security (2017) Using preventative radiological nuclear detection equipment for consequence management missions: operational job aids. DHS, New York City
10. United States Nuclear Regulatory Commission (2015) Implementation guidance for 10 CFR Part 37, physical protection of category 1 and category 2 quantities of radioactive material (NUREG 2155, Rev. 1). Government Printing Office, Washington DC
11. Wein L, Wilkins A, Baveja M, Flynn S (2006) Preventing the importation of illicit nuclear materials in shipping containers. Risk Anal 26(5):1377–1393
12. Zimmerman P, Loeb C (2004) Dirty bombs: the threat revisited. Defense Horiz 38(1):1–12

Chapter 14
Difficulties of Radiological and Nuclear Interdiction

Unfortunately, there is much more to radiological and nuclear interdiction than simply putting a large number of detectors into the field and waiting for results to come in. Radioactive sources can be shielded, and even sources of moderate activity will have a detection radius that represents only a tiny fraction of the area of a major city or an area to be searched.

Consider the 500 GBq Ir-192 source discussed earlier in this book; we calculated a detection radius of slightly less than 900 m. If we assume a detection radius of 1 km to make the calculations easier; the area in which this source (if unshielded) is detectable is about 3.14 km^2. The area of New York City is about 785 km^2; the area in which this source can be detected is a relative pinpoint on a map of New York City and those trying to locate the source are attempting to find that pinpoint. If the source is shielded or does not emit gamma radiation then the detection radius is even smaller and the source is correspondingly more difficult to find.

14.1 Challenges to Radiological Interdiction

Detecting an unshielded gamma-emitting source can be challenging, especially if the source has a relatively low activity, if it emits a low-energy photons, or if it emits a type of radiation (neutron, alpha, or beta) that has a very short range in air. But even higher-activity sources can be difficult to detect and/or to identify. Some factors that can make a source challenging to detect and identify include:

- The random nature of radioactive decay, natural radiation, and radiation detection
- The use of shielding
- Emission of alpha, beta, or neutron radiation
- Presence of nuclides with similar gamma energies.

P. A. Karam, *Radiological and Nuclear Terrorism*,
Advanced Sciences and Technologies for Security Applications,
https://doi.org/10.1007/978-3-030-69162-2_14

14.1.1 The Random Nature of Radioactive Decay

There are radiation instruments that will record every count received over a period of time; if the user sets it to count for one minute then, at the end of one minute, the display will show the total number of counts recorded from the time the "start" button was pushed until one minute has elapsed. So consider a technician whose job it is to push the button, record the number of counts, and then push the button again. Over time the technician will note that, while readings will vary from one button push to another, a distinct pattern emerges in which 68% of the button pushes result in count rates that are within one standard deviation of the actual background radiation count rate, 95% are within two standard deviations, and 99% are within three standard deviations. The problem is that a weak source, or a source that is at the limit of detectability will produce only a slight increase in radiation dose rates, and the instrument or the user must be able to determine when the very slight increase they see might represent a radioactive source and when it might simply be a statistical fluctuation in normal background radiation levels.

The difficulty of detecting a weak radioactive source

Pretend that you are searching for a radioactive source that might—or might not—be present. You know that, with the instrument you are using, the typical background radiation level is 100 counts per minute, or cpm (as shown in the standard distribution in Fig. 14.1). For the sake of this example (and to keep the numbers simple) we will assume that the standard deviation is 10 cpm—so one standard deviation (also known as one sigma) above background would be 110 cpm.

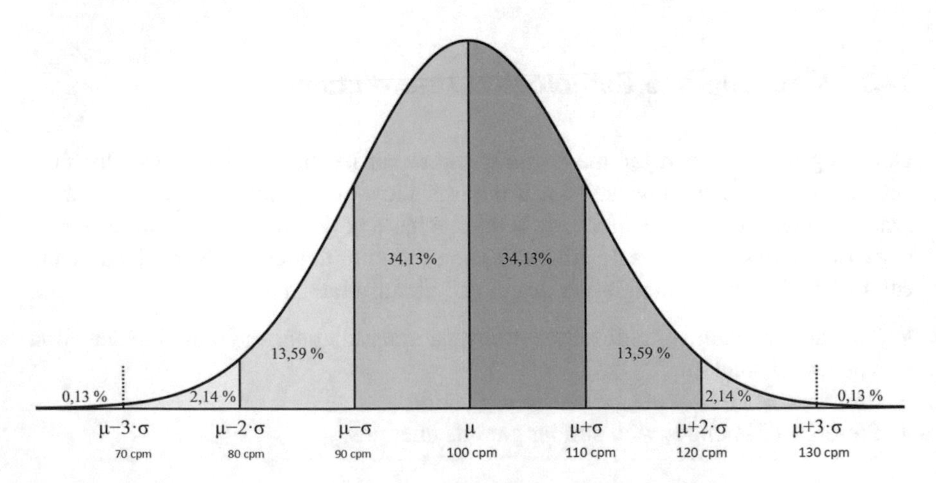

Fig. 14.1 The normal distribution (by Wolfgang Kowarschick and modified for this example by adding the count rate)

As you survey, after a fairly steady count rate of 100 cpm, you come to a location where the count rate jumps to 110 cpm. Should you report a possible detection?

So let's look at the normal distribution—we can see that about 34% of the time you should expect to see a reading between 100–110 cpm, simply due to normal fluctuations in the count rate from background radiation. So it's fairly likely that this elevated reading is due to the statistical fluctuation in normal background radiation levels and does not indicate the proximity of a radioactive source. In fact, most of the time a "detection" is not called away until radiation levels are at least three standard deviations above background, and under some circumstances the threshold for a detection might be as high as twice background levels (in this case, 200 cpm).

Now consider another example, a permanently mounted detector that records radiation levels every second—60 data points per minute and 3600 every hour. This means that a three-sigma fluctuation in background count rate (which happens about 0.13% of the time) will still produce 4–5 false alarms every hour, or about 100 false alarms daily. If we increase the tolerance to four sigma above background then about 0.05% of the counts will show up as alarms—about two false alarms per hour.

For a single person this is a manageable level, but what about for a network of, say, 25 detectors? If each detector is producing two false alarms each hour then the system as a whole will be alarming 50 times an hour, or more than 1000 times daily. So setting an alarm at the five sigma level can produce an unworkable number of false alarms with even a relatively modest number of detectors in the field. Yet, at the same time, setting alarms at the five-sigma level or higher can also lead to missing radioactive sources that are distant, well-shielded, or that have weak gamma emissions.

One way to help distinguish between these possibilities is to watch radiation levels carefully for a longer period of time to see if the slight increase drops back to near background or if it remains consistently elevated. If a source of radiation is actually present then the readings should fluctuate around a now-higher dose (or count) rate; if there is no source present then radiation levels should drop again after a short period of time and will continue to fluctuate around the expected background radiation levels.

Another way to try to make this determination would be to travel a short distance in one direction and observe the radiation levels, and then to repeat this in other directions. If levels increase while traveling in one direction and drop when traveling in the opposite direction then it is possible that a radiation source is causing these changes. However, the radiation source could also be a granite building, a brick building, a cemetery, or any of a number of other natural or non-threatening sources.

The person conducting the survey can also call for a large-volume scintillation detector coupled with a multi-channel analyzer to check to see if the slightly increased radiation levels are due to an increase in general background radiation dose rates or if the increase is in only one or two energies. An increase in background radiation levels will show up across the entire energy spectrum while an increase due to a single radioactive source will appear only in those energy channels associated with that nuclide—662 keV, for example, if due to Cs-137. Thus, a mobile radiation system might well identify the presence of a radionuclide (or, at least, might show a suggestive energy peak) that would escape a person reading a relatively simple PRD.

Having said this, the user should never lose track of the possibility that an apparent detection might not represent anything other than counting statistics, and the weaker the apparent "signal" the more likely that this is the case.

14.1.2 Shielding

Shielding can affect detection and identification in two ways: by attenuating the radiation emitted by a source and by changing the gamma energy spectrum as the radiation passes through the shielding. The former effect can impair detection, the latter can impair identification.

One important point to remember regarding radiation shielding is that, while it can reduce the "signature" of a radioactive source, it is hard to eliminate the radiation altogether. It is not difficult to reduce radiation dose rates or count rates by an order of magnitude or more but each additional tenth value layer adds additional weight and volume to the shield, making it easier to detect by virtue of its physical properties. For example, an industrial radiography camera that contains a 2 TBq Cs-137 source will weigh about 20–30 kg and will be about the size of a briefcase or a computer bag. But this source will still be detectable to a distance of several meters to anybody with a PRD, and to a much greater distance to a high-volume scintillation detector. To reduce dose rates further would require additional layers of shielding that would increase both the size and weight of the shield, with sources of higher activity the shielding required to hide the source from detection can weigh enough to affect the way a vehicle rides and handles; characteristics that police are accustomed to evaluating (Fig. 14.2).

Shielding can also introduce changes to a nuclide's spectrum, primarily through a phenomenon known as Compton scattering. For example, consider a Cs-137 gamma ray photon that is absorbed and re-emitted in a random direction by an electron in the radiation shielding. When the photon is emitted it will have a slightly lower energy than the original photon; perhaps it is emitted with an energy of, say, 650 keV instead of the original 662 keV. This scattered photon can, itself, be absorbed and re-emitted within the shield, and again and again—each time having a slightly lower energy. As these photons continue being scattered and emerge from the shield the original gamma peak will begin to "smear out" into lower energies. As additional shielding is added the gamma peaks become less distinct, and the software that identifies radionuclides based on gamma energy finds it increasingly difficult to identify the radionuclides with a high degree of certainty.

Fig. 14.2 Three older-style industrial radiography "cameras." Each of these is a lead or depleted uranium radiation shield containing a radioactive source of a few TBq in activity. (Author's photo)

However, regardless of Compton scattering, it is worth noting that the spectrum will not completely block every gamma photon emitted and some of these gammas will emerge from the shield with the original peak energy. In the case discussed here, while the peak at 662 keV might become indistinct and hard to identify, above that energy there will only be the normal background energy spectrum. Identifying this "cutoff" energy can help to identify the radionuclide present, although, that being said, this is more than what can reasonably be expected of an emergency responder of even of many radiation safety professionals, although it would be within the capabilities of a Tier 1 reachback facility.

In addition, gamma photons with energies higher than 1.022 MeV will produce electron-positron pairs when passing through the shield; these recombine and annihilate each other (a positron is an antimatter electron), generating 511 keV annihilation radiation. This annihilation peak will be evident in the spectrum; in and of itself this is not sufficient to identify, say, Co-60, although it can suggest its presence. However, it could also indicate the presence of a positron-emitting radionuclide (e.g. the F-18 used for PET scans), Sr-90 (which decays to Y-90, which emits a 2.2 MeV gamma), or other similar radionuclides. And, even if the presence of an annihilation peak leaves a number of possible radionuclides that could be present, it also eliminates all radionuclides that emit neither positrons nor high-energy photons, which can help limit the field of possibilities significantly.

14.1.3 Emission of Alpha, Beta, or Neutron Radiation

The majority of radiation detectors used for interdiction are gamma scintillation detectors that do not detect alpha, beta, or neutron radiation. This makes sense because neutron-emitting sources are rare and because of the relatively short range of alpha and beta radiation in air; while gamma radiation can be detected up to several kilometers from a sufficiently powerful source, alpha radiation has a range of only a few millimeters, beta radiation can travel only 6 or 7 m, and neutron radiation can travel up to several hundred meters in air only if the source is sufficiently high-activity. Any or all of these types of radiation can be difficult to detect with the types of radiation detectors that are most commonly used for interdiction.

Many alpha- and beta-emitting radionuclides also emit gamma radiation, or gamma radiation is emitted by their progeny. Cesium-137, for example, does not emit a gamma; Cs-137 is a pure beta-emitting radionuclide. But Cs-137 decays to form Ba-137 m, and Ba-137 m emits gamma radiation with a half-life of only about 6 min. Thus, while a gamma detector will not detect Cs-137, it is easily able to detect its Ba-137 m progeny. Similarly, while Sr-90 emits only beta radiation, the high-energy gamma emitted by its Y-90 progeny is easily detected. Similarly, alpha-emitting Am-241 and Ra-226 also emit distinctive gammas, as do many other beta and gamma-emitting radionuclides.

In addition, beta radiation, when passing through material, will sometimes undergo interactions that result in the emission of x-ray radiation called bremsstrahlung. Even though it is very difficult to identify a radionuclide by the bremsstrahlung emitted, its very presence will serve to increase radiation levels, alerting the user to the presence of radioactivity, even if it cannot be identified.

Neutron radiation can be more problematic; there are very few radionuclides that undergo spontaneous fission and emit neutrons (Cf-252 is the most common). Most neutron sources use a combination of Am-241 or another alpha-emitting nuclide in conjunction with beryllium; when the alpha particle interacts with the beryllium a neutron is ejected. Thus, many neutron sources will also emit the characteristic 60 keV gamma of the Am-241, in addition to gamma radiation emitted when the neutron is absorbed by a target atom. As with bremsstrahlung, the gamma radiation might not be readily identifiable, but it can serve to increase the general radiation dose rates. In addition, many departments tasked with interdiction are equipped with neutron detectors, increasing the likelihood of interdicting neutron-emitting materials.

14.1.4 Presence of Nuclides with Similar Gamma Energies, Limited Gamma Libraries

While no two gamma-emitting radionuclides give off gammas with identical energies, there are some radionuclides that emit gamma rays with very similar energies, and this can lead to mis-identification, especially when using scintillation detectors to perform the nuclide identification. The reason for this is that scintillation-type detectors—especially sodium iodide and cesium iodide—have a relatively low energy resolution compared to high-purity germanium, making it difficult to resolve and identify closely spaced gamma energy peaks. Figure 14.3 shows the same gamma energy spectrum as measured by several different detectors—looking at these peaks, it is

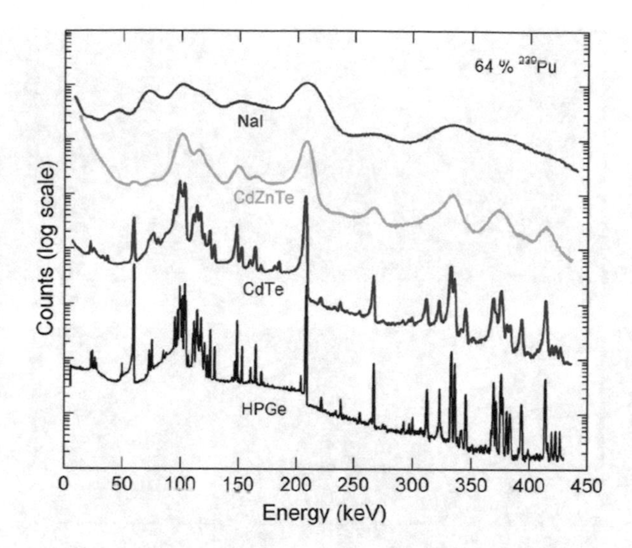

Fig. 14.3 Comparing the energy resolution of various gamma detectors. Sodium iodide (the top line) is inexpensive and operates at room temperature, but has a low energy resolution. High-purity Germanium (HPGe) has outstanding energy resolution but is very expensive and must be cooled using liquid nitrogen or with a refrigeration unit. It is easy to see how a sodium iodide detector could miss identifying closely spaced gammas. (From Basics of Gamma Ray Detection, Los Alamos National Laboratory document LA-UR-17-28266 [7])

easy to see how the ubiquitous sodium iodide (NaI) scintillators would lump together closely spaced energy peaks that would be individually resolved by a high-purity germanium (HPGe) detector.

Perhaps the best-known such gamma peaks are those of Ra-226 and U-235, at 186 keV and 185 keV, respectively. Sodium iodide and cesium iodide detectors are unable to resolve these two peaks, and even HPGe can do so only with difficulty. Since Ra-226 is commonly found in naturally occurring radioactive materials, including granite, uranium minerals, and wastes from oil and natural gas production and processing, and U-235 is both naturally occurring in low quantities (0.72% of natural uranium) and is used in nuclear weapons, it is not uncommon for a scintillation-type RIID to mistakenly identify Ra-226 as "special nuclear materials" (SNM). This is shown in the HPGe spectra shown in Fig. 14.4. Other nuclide pairs with closely spaced gamma peaks include I-131/Ba-133 (364/356 keV), I-129/I-125 (35/40 keV), and Na-22/F-18/Ga-68 and a handful of additional nuclides that all emit positrons with the corresponding 511 keV annihilation gamma.

In addition to these closely spaced gamma peaks, some scintillation detectors, especially older detectors that are not energy-stabilized, can exhibit "energy drift" as temperatures change. This can cause the mis-identification of gamma peaks or can lead to a failure to properly identify a gamma peak. As one example, the author participated in a gamma survey in which, due to changing detector temperatures, the K-40 gamma (at 1.46 MeV) was identified as Co-60 (1.33 MeV); that this was an

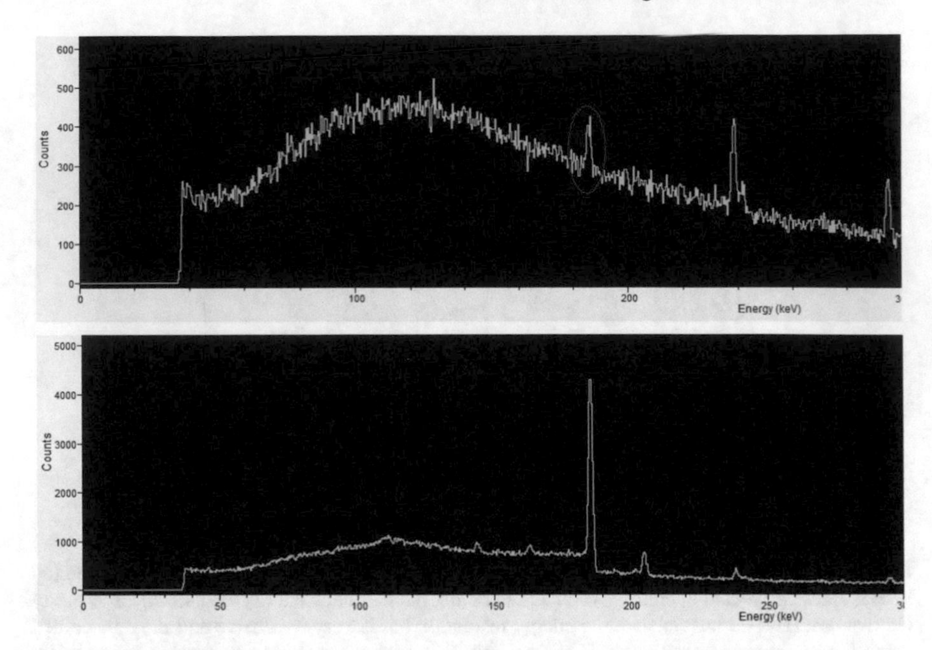

Fig. 14.4 Natural background spectrum (top) and U-235 (bottom) as measured using HPGe. In the top spectrum, notice the close energy spacing of the U-235 and Ra-226 lines (circled) while, in the bottom spectrum we can see that the U-235 source swamps the natural Ra-226. It is easy to see how even a high-resolution detector can fail to resolve these two radionuclides. Spectra obtained by author

error was clear during a visual examination of the spectrum, which was confirmed by briefly turning the instrument off, noting that the "ID" disappeared when it restarted. Errors of this magnitude are rare, especially with newer instruments, but they can occur.

Somewhat more common is to have a gamma peak that cannot be identified because the instrument's nuclide library does not include that particular peak or the associated radionuclide. No single instrument maintains a comprehensive library of every gamma energy of gamma-emitting radionuclide; such a library would be unwieldy, and would be bound to contain a huge number of radionuclides that are highly unlikely to be seen in the field due to having a very short half-life, being rare synthetic nuclides (e.g. Np-237, Pu-240, or Es-254), nuclides produced only under very specific circumstances (e.g. in high-energy particle accelerators or in nuclear weapons debris), and so forth. This means that, every so often, there will be a gamma peak that cannot be identified by the instrument and that might need to be sent to a national laboratory (in the US) or to another advanced facility for identification via reachback.

14.2 Complications with Nuclear Interdiction

One of the most significant challenges to nuclear interdiction is the fact that nuclear weapons do not necessarily emit high levels of radiation and are not unequivocally detectable at anything other than relatively short distances. While specific information about most nuclear weapons tends to be classified, some scientists ([5] and [6]) were permitted to make a series of measurements in close proximity to Soviet SS-N-12 "Sandbox" nuclear-armed missiles using both U-235 and Pu-239 as the fissile material. These measurements indicated that the weapons surveyed would have had a maximum detection range of about 4–5 m using the instruments available at that time.

Fetter's paper was written in 1990, using both Russian and American radiation detectors of that era and it is natural to wonder if the results and conclusions of this paper are applicable today. It seems reasonable to conclude that they are for reasons enumerated below:

- While the radiation instruments used today are considerably different from those in use in 1990, the detectors themselves have not changed much; many of the advances in the design of the most common radiation instruments involve the user interface, adding digital (versus analog) processing and displays, and developing more powerful software for identifying analyzing gamma peaks. However, the most commonly used detectors continue to use sodium iodide, high-purity germanium, and a variety of neutron detectors—the same materials used in 1990.
- The instruments used by emergency responders are not the most-advanced instruments that have been developed over the last few decades; the most-advanced instruments continue to be used by national laboratories, military, and intelligence communities. Instruments used by emergency responders are rugged and reliable but they are not cutting-edge.
- The weapons surveyed are reported to have had a yield of over 300 kt [3] which is considerably higher than the anticipated yield of a terrorist group's improvised nuclear device. This means that the weapons surveyed almost certainly contained much more U-235 and Pu-239 than would a terrorist device and, if anything, would be detectable at a greater distance than an anticipated IND.
- The detection method described in this paper involved detecting specific gamma energies, which can be detected at a greater distance than less sophisticated means (e.g. gross count rate, gross radiation dose rate) because narrow energy bands have a lower background count rate than the energy spectrum as a whole.

For these reasons, it seems reasonable to conclude that the results of this paper remain relevant today and that a reasonable maximum detection radius for a terrorist IND is less than 10 m. Thus, nuclear weapons are even more difficult to detect during interdiction operations than are radiological devices.

Nuclear weapons also emit neutron radiation from the spontaneous fission of heavy atoms, from an alpha-neutron interaction in which an alpha particle is absorbed by a heavy atom, prompting the release of a neutron, and from the occasional cosmic ray-induced reaction that releases a neutron. According to Ogawa [8] a 4-kg mass of weapons-grade plutonium (containing no more than 6% Pu-240 by weight) will emit about 400,000 neutrons per second and will have a detection distance of about 12 m when depending on gamma detection and 25 m when identifying the device using neutrons. Fetter et al. [4] also note that highly enriched uranium emits far fewer neutrons than Pu-239; 12 kg of HEU will emit about 1400 neutrons per second with a correspondingly lower detection radius (about 1.5–2 m). Fetter et al. also look at photon emissions, concluding that gamma emissions will be indistinguishable from background at a distance of about 3-6 m from 4 kg of plutonium and as far as 8 m from 12 kg of HEU. Given that a working weapon will require a higher mass of either of these nuclides the gamma and neutron emissions will be higher and the detection radius correspondingly larger, but likely no more than double the distances noted here.

Another challenge to interdicting an IND involves identifying the presence of special nuclear materials (i.e. weapons-grade uranium and plutonium). In particular, the identification of U-235 and Pu-239, while never easy, can be made more difficult by the presence of other radionuclides. The most common interference comes from Ra-226, which has a gamma energy nearly identical to that of U-235—close enough, in fact, that it is not uncommon for any brand of RIID to mistakenly identify Ra-226 as "SNM" or as U-235.

While this sort of mistaken identity can be annoying, there are other means by which isotopes can be purposely masked in order to hide their identify. In a technical document, for example, the US Government Accountability Office notes that the presence of SNM can be concealed using "masking" radionuclides, making the detection and identification of highly enriched uranium, weapons grade plutonium, and other nuclides of interest more difficult [9]. However, it should also be noted that, while it is possible to mask the presence of nuclear materials, doing so requires the use of sufficient quantities of licensed radioactive materials. Thus, a group trying to mask a nuclear device would have to do so by either stealing or obtaining a license to purchase other radionuclides. This might be an effective stratagem, but it also adds time, complexity, and the chance of being caught to a plot that is likely already fairly complex.

14.3 Instruments and Their Maintenance

Radiation instruments are a crucial part of any interdiction efforts, but they are even more important in nuclear interdiction than for radiological. One reason for this is the low radiation "signature" mentioned earlier in this chapter; detecting and identifying a weak radiation signal is far more difficult than doing so for a strong one. Adding to the difficulty is the relatively low energy of the majority of the gamma radiation emitted by uranium and plutonium and the difficulty of surveying for both these gammas as well as challenges associated with neutron detection.

What this means is that any attempt to interdict a nuclear weapon—improvised or otherwise—will be heavily dependent on having appropriate radiation detection instruments and using them properly, as well as properly maintaining them.

Given the relatively weak signal from a uranium weapon, large-volume gamma detectors are most likely to be successful in detecting a nuclear weapon, if only because alpha and beta radiation have such a short range and the neutron signature is so small. A plutonium weapon might be easier to detect via both gamma and neutron emissions; however the need for a more complex implosion design for plutonium weapons makes this design less likely for a terrorist group fabricating a weapon [10].

Maintaining one's instruments is as important as is having the correct detectors, and any agency that is purchasing radiation instruments must include these costs in their budget in addition to the costs of instrument replacement and operator training. Instrument maintenance includes both preventative and corrective maintenance as well as initial and periodic calibrations to ensure that the meaning of various instrument readings is clear and can be interpreted properly. Consider, for example, the problem discussed in Sect. 12.3, in which instrument readings were used to help to quantify the activity of a number of interdicted sources. In order for such calculations to be accurate, there must be a way for the person performing the calculations to know the detection efficiency of their radiation detectors for the nuclide(s) in question at a variety of distances—the counting efficiency at a distance of, say, 10 m will be very different from that at an altitude of 150 m or more due to the increased distance as well as to the attenuating effects of 150 m of air. For this reason, it might be necessary to calibrate airborne instruments by making repeated measurements of the same source or the same area at a variety of altitudes, comparing airborne readings with those made on the ground using properly calibrated instruments.

In addition to performing these initial calibrations, instruments that are used regularly should also be checked or calibrated periodically to ensure they are operating properly. One reason for this is to ensure the instrument is responding properly to radiation, to ensure that alarms sound at the proper radiation dose rate, and (in some jurisdictions) for evidentiary reasons as well; to be able to demonstrate to a judge and/or jury (for example) that elevated radiation readings are most likely the result of the presence of radioactive materials rather than a malfunctioning instrument.

There are several levels of periodic checks that should be performed on radiation instruments with varying intervals:

- **Bump checks** use a low-activity radioactive source to confirm that the radiation detector responds to the presence of radiation. This need not be a calibrated or characterized source, it only need be capable of causing the instrument to respond. To check a sodium iodide (gamma) detector, one may use a low-activity (a few tens of kBq) gamma source, a radioactive mineral sample, thoriated welding electrodes, and so forth (a neutron instrument will require a neutron source), provided the source used emits enough radiation to cause the meter to deflect or to register more than a few standard deviations higher than normal background levels. Bump checks should be done daily or weekly when an instrument is in use.

- **Field response tests** use a radioactive source of known strength to confirm that the radiation instrument produces a reading that is within a specified percentage of the expected reading. For example, a Cs-137 source that is known to give a reading of 15 μGy hr^{-1} when held at a specified distance from the detector should reliably produce this reading (within acceptable variability) whenever tested. A field response test can also include checking to confirm that the displayed background radiation dose rates are within acceptable limits. The field response test is not the same as performing an instrument calibration, but it can verify that instrument readings are reasonably accurate. Field response tests should be performed annually or semi-annually.
- **Calibration** is a more detailed process [1] that involves confirming the instrument's physical condition, testing alarms, checking electrical and electronic systems as well as confirming the instrument's response to multiple radiation dose rates and/or count rates; if the instrument does not respond within acceptable margins then the calibration procedure also includes adjusting the instrument as necessary. Calibration of instruments used for health and safety or regulatory compliance purposes is typically performed annually; for instruments used only for interdiction or for informational purposes calibration is recommended when the instrument fails its bump or field response checks or every three to five years, whichever is more frequent.

Instrument calibration can be expensive, costing thousands, tens of thousands, or even hundreds of thousands of dollars annually depending on the number of instruments an agency has purchased. In addition to the cost of calibration, the agency must also collect the instruments, ship them to the calibration facility, and the instruments cannot be used for up to several weeks—for these reasons it is not uncommon for agencies to neglect calibration altogether. However, if field response checks show that an instrument is indicating properly, it might not be necessary to conduct annual calibrations; calibrating every 3–5 years (or when an instrument fails a field response check) can reduce costs while still ensuring that an instrument is operating properly.

In addition, users should perform regular checks on their instruments to confirm that the instruments are not visibly damaged, to replace depleted batteries, to confirm the instrument's calibration status, and to ensure that it is the proper instrument to use for the mission to be performed.

14.4 Managing Large Interdiction Efforts

Consider the commander of an interdiction mission—the person responsible for managing and coordinating the efforts of five hundred people on foot, in vehicles, in the air, and on the water. Four hundred of these personnel are likely carrying

PRDs, a few dozen will be equipped with backpack detectors, a dozen Tier 2 responders might have hand-held RIIDs, and another dozen will be in vehicles (automobiles, helicopters, and boats) carrying large-volume scintillation detectors. The remaining 40–50 personnel will be staffing portal monitors, monitoring communications, dispatching Tier 2 responders to help adjudicate alarms, running replacement detectors to fixed locations, monitoring data and mapping displays, and all of the other minor tasks that help those in the field to work effectively. In addition, for a major event, the State and Federal governments (in the US) might send personnel and equipment as well, which will need to be coordinated with the local resources to minimize overlapping efforts.

A single interdiction effort might stretch across an entire city if the event being covered is a marathon or a large parade, it might last for days or weeks if it is to cover an extended event such as a sport championship series or a major international conference, it might be limited to a relatively small area for a very controversial event, or any of a number of other possible situations. Every scenario will call for a different level of effort and a unique disposition of the available resources aimed at ensuring the security of a particular event in a specific location and that takes place in the social and political climate of that time.

One of the key stations during an interdiction is the data display—a single location showing the readings from all of the instruments, fixed and mobile, deployed in the field. The person(s) monitoring this display might be the only person with a coherent view of radiological conditions through the monitored area, who can not only review alarms as they occur, but who can also see which areas might require additional resources or more careful attention. The subject matter expert (SME) will likely not have the authority to dispatch additional resources as needed; if not then the SME will need to make recommendations to the event commander or other command staff as appropriate, who can give the appropriate directions to personnel in the field (Fig. 14.5).

For an event of any complexity the commander will be well-advised to appoint a command staff to attend to the details while the commander maintains a high-level view of the entire event, similar to that which is established during response to fires and similar emergencies [2]. This command staff might include a radiological expert to interpret the data as it is collected and transmitted from field locations and to act as a subject matter expert, a health and safety officer (especially on very cold or hot days, and particularly when personnel are wearing body armor), a logistics officer to ensure that necessary personnel and equipment are available, liaisons with other levels of government, and so forth. In addition to helping the commander maintain an appropriate span of control, in the event that an attack takes place this will also provide the basis for the incident command structure that will be needed to manage the initial emergency response efforts. Figure 14.6 shows one example of such a command structure.

Fig. 14.5 An example of a radiological interdiction map display from an actual interdiction mission, such as would be maintained by the event commander. The data on this map was transmitted from a combination of mobile systems, backpack units, and handheld RIIDs

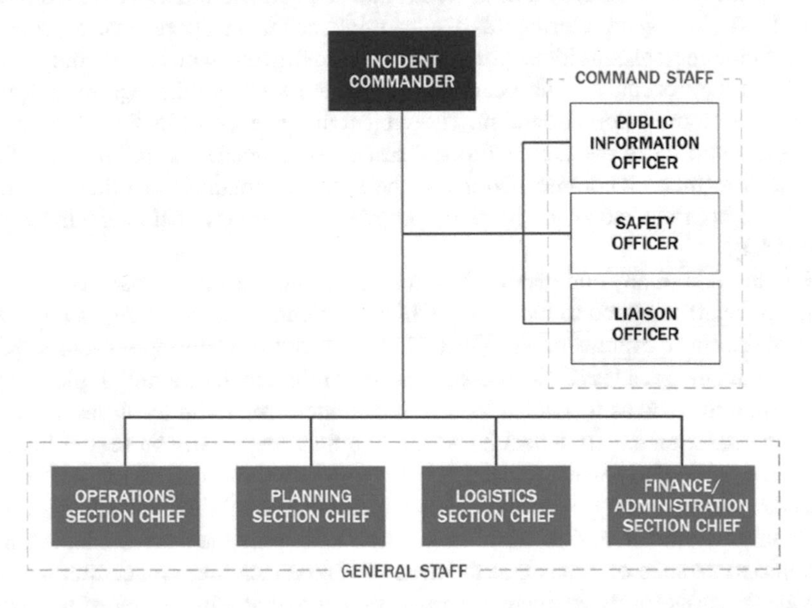

Fig. 14.6 An example NIMS command structure with a single Incident Commander. FEMA [2]. A more complex event requiring multiple agencies might have a unified command consisting of representatives of each major agency present in place of the Incident Commander

References

1. American National Standards Institute (2013) N323AB: radiation protection instrumentation test and calibration. ANSI New York City
2. Federal Emergency Management Agency (2017) National incident management system, 3rd ed. FEMA Washington DC
3. Federation of American Scientists. Weapons of mass destruction: SS-N-12 sandbox. Available online at https://fas.org/nuke/guide/russia/theater/ss-n-12.htm. Accessed 18 Nov 2020
4. Fetter S, Prilutskii O, Rodionov S (1989) In: Altmann J et al (eds) Passive detection of nuclear warheads in verification of arms reductions. Springer, Berlin
5. Fetter S, Cochran T, Grodzins H, Zucker M (1990) Gamma-ray measurement of a soviet cruise-missile warhead. Science 248:828–834
6. Fetter S, Frolov V, Miller M, Mozley R, Prilutsky O, Rodionov S, Sagdeev R. Detecting nuclear warheads. Sci Glob Secur 1:225–253
7. Stinnett J and Venkataraman R. (2017) Basics of Gamma Ray Detection.
8. Ogawa I (1990) Detection of nuclear warheads by radiation: a brief review. In: Miyajima M, Sasaki S, Doke T (eds) Nuclear disarmament, safeguards, and physical protection. National laboratory for high energy physics, Tsukuba, Ibaraki (Japan). Workshop on radiation detectors and their uses; Tsukuba, Ibaraki (Japan); 30–31 Jan https://inis.iaea.org/search/search.aspx?orig_q=RN:22067133. Accessed 18 Nov 2020
9. US Government Accountability Office (2009) Combating nuclear smuggling: DHS improved testing of advanced radiation detection portal monitors, but preliminary results show limits of the new technology. GAO, Washington DC
10. Zimmerman P, Lewis J (2006) The bomb in the Backyard. Foreign Policy. November/December, pp 32–39

Chapter 15
Initial Emergency Response Efforts

The initial moments of any emergency are rushed, confusing, and essential; how these moments are handled can set the stage for how the emergency itself will be handled, and mistakes or oversights at this early stage can escalate and propagate forward if they are not recognized or corrected. Thus, it is important that key personnel—the Incident Commander, Health and Safety Officer, and others—be able to attend to not only the "routine" risks and issues that accompany any emergency, but the radiological matters as well. Further, those engaged in the response must also be able to properly prioritize both radiological and non-radiological concerns rather than fixating on radiological matters because they are novel and frightening, or ignoring them because they are unfamiliar and poorly understood.

While there has never been a radiological attack there have been many radiological accidents and incidents that we have learned from. To that we can add the experience from nuclear reactor accidents in the United States, the Soviet Union, the former Yugoslavia, and Japan as well as the further experience from the atomic bombings in Japan (although those will not be the subject of this chapter). Finally, people have been working with radiation and radioactivity for more than a century and have learned to do so safely and to manage the risks it poses in industrial, reactor, and emergency settings; the lessons that have been learned in these decades of experience can be applied to many aspects of radiological and nuclear emergency response.

15.1 The First 100 Min

In 2017 the US Department of Homeland Security published a document detailing how they anticipated the first 100 min of a radiological response might unfold [12]. And, while they acknowledged that, in an actual emergency, events might unfold differently or might take more time, this document laid out the important tasks that are expected to take place during the first stages of a response to a radiological

P. A. Karam, *Radiological and Nuclear Terrorism*,
Advanced Sciences and Technologies for Security Applications,
https://doi.org/10.1007/978-3-030-69162-2_15

event. The "First 100 mins" plan is science-based, but was developed with input from emergency responders from across the United States. The DHS anticipates that the several tasks will need to be accomplished quickly:

1. Recognizing that a radiological attack has taken place
2. Informing appropriate agencies of the event and transmitting initial information from the scene
3. Initiating life-saving operations and securing and managing the scene
4. Measuring and mapping radiation levels
5. Evacuating and monitoring members of the public.

Each of these tasks is, itself, comprised of one or more "tactics" used to help accomplish the task. Some of these tasks will be described briefly here, others will be discussed in greater depth in the subsequent sections of this chapter.

15.1.1 Recognizing that a Radiological Attack Has Taken Place

It might not be immediately obvious that an attack has taken place. A covert attack, for example, is not announced with an explosion and it might not be obvious that an attack has even been launched; Alexander Litvenenko was administered Po-210, but the fact that he had been poisoned with radioactivity was not recognized for nearly a month, until the day before his death. But even an overt attack—a bomb—might not be recognized as a "dirty bomb" unless some of the responders are using radiation detectors. In some cities, police and fire vehicles carry radiation detectors and elevated radiation levels would be noticed as soon as the first vehicle rolls up at the scene; in other cities this is not the case.

No matter when the first radiation detector arrives on the scene, it still might not be clear that a radiological attack has occurred. Consider, for example, a police officer who rolls up at the scene of an explosion and his radiation detector alarms at a radiation dose rate of, say, 0.50 μGy hr^{-1}. Should the officer call away a radiological emergency?

The answer is not as clear-cut as it might seem. For example:

- Is the radiation detector calibrated?
- What is the normal background radiation dose rate, not just in the city but in the area of the explosion?
- Is there anything in the area that could cause the detector to read high (hospital, veterinary clinic, granite structure(s), radio or radar transmitters, industrial radiography, and so forth)?

One approach to "calling away" a radiological event is to require instrument readings to be a multiple of normal background readings (such a procedure, for example, might call for an instrument reading of two or three times normal background radiation dose rates as indicating a radiological event has taken place). This is the standard approach to determining the presence of contamination at a nuclear power plant and other radiological or nuclear facilities. One problem with this approach, however, is that it needs to be implemented by personnel who have a fairly detailed understanding of their instruments and what they normally read at their facility; in a large city with variable background radiation levels, measured with a variety of instruments used by personnel who are not full-time radiation safety workers such an approach might be challenging to implement.

Different types of radiation detectors, for example, can have different readings in the same location; a non-energy-compensated GM, for example, can easily have a reading that is double the reading of a pressurized ionization chamber, both of which might read differently than a PRD using a sodium iodide detector. While a radiation safety professional would have the experience and training to understand which of these instruments should be considered to be providing a more reliable "background" reading, most emergency responders lack this training and might easily not use an appropriate "background" reading to serve as the basis for determining that a radiological release has occurred.

Consider, for example, a city in which typical background radiation dose rates are between 0.05–0.010 μGy hr^{-1} as measured with a pressurized ionization chamber. A PRD that uses a cesium iodide crystal will tend to have higher readings, so a police officer performing a survey using such an instrument might routinely get readings of 0.10–0.15 μGy hr^{-1} in these same areas. Which readings should the officer use to determine "background" radiation dose rates and what level represents a radiological release? One could make a good argument that double "background" readings might be as low as 0.10 μGy hr^{-1} or as high as 0.30 μGy hr^{-1}, depending on the instruments used. Further, radiation dose rates next to a granite building or wall might be twice as high as the measured background radiation dose rates elsewhere in the city. Lacking a detailed understanding of the differences between various types of radiation instruments and the manner in which background readings can vary from place to place in their city, it is entirely possible that a police officer who turns their instrument on at the beginning of the day, noting a background reading of, say, 0.1 μGy hr^{-1} at their precinct station might later find him or herself responding to a small explosion near the granite wall mentioned earlier with a reading of 0.3 μGy hr^{-1}. Should this be reported as evidence of a radiological attack?

A different approach would be to conduct a thorough survey of the city, noting the variations in radiation dose rates from place to place and determining not only the typical radiation dose rates, but also the variability across the city, including "hot spots" such as the granite wall noted above—and as measured with a number of different instruments. These readings can then form the basis for determining the radiation dose rate at which a radiological incident would be declared; given the readings noted above, it would be reasonable to consider developing a city-wide procedure under which any dose rate in excess of 1 μGy hr^{-1} as measured by any dose rate instrument would be considered evidence of a radiological incident, with the circumstances and (possibly) additional information used to determine whether or not the incident represented an accident, the routine use of radiation or radioactivity, or an attack of some sort.

Table 15.1 PRD alarm adjudication examples

Reading (μGy hr^{-1})	Detector	Circumstances	Adjudication
11	CsI PRD	500 m from construction site with industrial radiography taking place	Innocent alarm
1.3	Energy-Comp. GM	Elevated dose rates are distributed over an area of about 2 ha	Covert attack
2.8	NaI PRD	In vicinity of nuclear medicine clinic	Innocent alarm
5.7	Pressurized ionization chamber	Near truck with soil density gauge in bed	Innocent alarm
0.2	NaI PRD	At scene of pipe bomb explosion	Possible attack, possibly non-radiological
0.5-1.8 Cs-137, I-133 nuclide ID	NaI mobile system with high-volume detectors	Site area emergency at nearby nuclear power plant	Possible reactor accident

Some examples of various circumstances are summarized in the following Table 15.1.

Of the events noted above, two likely represent terrorist attacks (one radiological and one non-radiological in nature), one represents a likely reactor accident (due to the presence of volatile fission products), and three are likely innocent alarms.

The DHS "100 minute" document also recommends "*After an initial indication that radiation is present, first responders take at least two readings, in at least two locations, with at least two separate radiation detection instruments to confirm that elevated radiation levels above background are present at the…scene*" [12]. This document estimates that recognizing that radiation is present will require no more than five minutes for agencies that bring radiation instruments to the scene.

15.1.2 Inform Agencies and the Public of the Event and Transmit Information from the Site

As soon as an event is determined to be radiological in nature notifications must be transmitted to appropriate agencies. One reason to make these notifications is to inform the appropriate levels of government of an event of which they should be aware; any radiological attack is likely to represent an attack against the nation (even if confined to a relatively local area) that might require a national response to adequately resolve. In addition, if the event represents a terrorist attack then it will be necessary for the national government to quickly identify those responsible so that an appropriate national response can be formulated; the faster the national government becomes aware of the fact of an attack, the faster they can respond as appropriate. Of more importance to those in the city that was attacked, these notifications will help to determine the magnitude of the event, the need for a local and/or national response, and will help the relevant agencies at all levels of government to coordinate their actions in response to the event. In addition, information transmitted from the scene will help the government to understand the geographical scope and radiological severity of the event and the appropriate precautions (e.g. shelter-in-place, evacuation, etc.) that are required to see to public health and safety.

The initial information from the scene will include announcing the fact that the event is radiological in nature. This information will, in turn, set in motion the other response and recovery actions that will take place over the subsequent days, months…even years of work.

Once the emergency response agencies and the elected and appointed officials have been informed that a radiological attack has taken place they should inform the public of this fact and of the recommended actions that should be taken to protect their health and safety. In most cases it will make sense to instruct the public to take shelter immediately in the nearest building that is safe and structurally sound, to shut any open doors and windows, and to listen to the news, with more targeted and more precise instructions to follow as more is known. In some cities there has been debate over what to tell residents—in particular, at what point to first use the word "radiation." One approach is to inform the public immediately of the radiological nature of the event, although many have expressed concerns that this could frighten the public and cause them to take actions that are inappropriate. For example, in the aftermath of the nuclear reactor accident at Three Mile Island there were far more people who self-evacuated than was called for [11]. Harrisburg Pennsylvania is a relatively small city; in 2020 the city itself had a population of nearly 50,000 with a population of about ten times that high in the greater metropolitan area. Nearly 45,000 people evacuated the Harrisburg metropolitan area in the aftermath of the Three Mile Island accident; a comparable flight from New York City would be over 1 million people, with the concomitant risks associated with this number of frightened drivers and their passengers driving away from an event that frightens them immeasurably. For this reason, there are those who advise delaying the mention of radiation, feeling that, by taking some time to prepare the public for having been struck with a radiological attack, they might behave more rationally.

A competing approach discussed in meetings attended by the author would be to lead the public gradually into the fact that a radiological attack has occurred. In this approach, the initial report might state that a bomb had gone off, followed by a report of hazardous materials found at the scene, and finally noting that radioactivity had been identified and measured. The thinking behind this approach is that the slower build-up to announcing a radiological release might reduce the fear on the part of the public, hopefully leading them to respond more thoughtfully and to pay attention to governmental recommendations. The chief potential concern with this approach is that there will undoubtedly be members of the media present at the scene within a short time; the media might or might not have radiation detectors of their own, but they are likely to notice a number of emergency responders performing radiation surveys and wearing PPE. If the government is perceived to be deceiving the people in the first moments of the emergency response it will be difficult to recover the public's trust, making subsequent orders and recommendations less effective than would otherwise be the case. And even if the government does not exactly lie to the public (radioactivity is "hazardous material" after all), the distinction between lying and being disingenuous is likely to be lost on the public during a period of high stress.

The other part of the "inform" mission is to inform other levels of government and appropriate governmental agencies of the event and the need for assistance. In the United States, for example, the Federal government does not automatically provide assistance—they must be asked to do so by the state that was attacked. Although this particular task might seem like a rather odd formality, it remains important nevertheless. Consider, for example, the Health Commissioner hearing the news of a radiological attack in their city. The report might well be accurate; it might also simply be the result of a reporter showing up at the scene of an explosion, hearing radiation detectors registering background radiation and, not understanding that background radiation exists, reporting that a radiological attack took place. The Health Commissioner cannot spring into action based on radio or internet news reports, neither can the Federal government send assistance based on such anecdotal information; there has to be a formal notification through a recognized communication channel before the Health Commissioner (and Commissioners of other agencies) can order their resources to respond. For similar reasons, there must be an official notification to other levels of government to ask for their resources to be sent to the scene to assist. As an aside, this means that the city must have a plan as to how these resources will be incorporated into the city's emergency response efforts; it can be useful for the various agencies at the different levels of government to have such discussions before an attack or other form of emergency occurs so that all parties understand their role, where to report, and can work out the logistics associated with their response.

15.1.3 Measuring and Mapping Radiation Levels

One of the crucial initial tasks will involve measuring and mapping radiation levels at the scene and throughout the city; in particular it will be necessary to quickly determine the direction and "footprint" of the radioactive plume so that radiological boundaries can be established and so that members of the public (and emergency responders) will know which areas require taking appropriate radiation safety precautions upon entry. Immediately upon recognizing a radiological attack the Incident Commander might choose to establish default Hot Zone boundaries; one recommendation [3, 6] is to establish default boundaries at a distance of 250 m in all directions around the site of the explosion and extending them 2 km downwind. This should encompass the areas with high levels of radiation as well as the majority of the contaminated areas. As time goes on, these initial boundaries can be adjusted as necessary to reflect new measurements. As a general rule of thumb, there

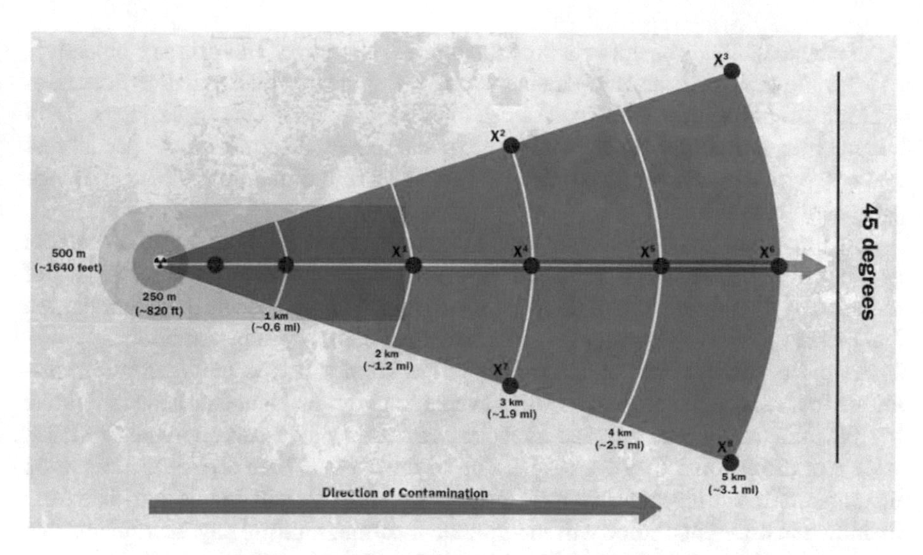

Fig. 15.1 An example preliminary 10-point survey pattern from the "First 100 Minutes" document [12]. This survey can be completed quickly and will provide an initial idea of the direction, size, and contamination levels of the plume

are advantages to setting initial boundaries generously and then collapsing them inwards as more information becomes available; expanding boundaries outwards during the source of an emergency response can raise concerns about contamination and personal exposure among both responders and members of the public in the zones that were initially outside of the Hot Zone boundary (Fig. 15.1).

Preliminary mapping should begin fairly quickly, if only to be able to establish the Hot Zone boundaries and to delineate the footprint of the radioactive plume, which can extend for several kilometers downwind from the site of the explosion. Finding the broad outlines of the plume can be accomplished reasonably quickly with the proper equipment and the results of this can be used to modify the initial shelter-in-place instructions.

The "First 100 min" document recommends the Incident Commander assemble three teams to conduct the initial radiological characterization and mapping, using a wedge-shaped pattern extending downwind from the point of the explosion. The first quick survey can roughly determine the plume footprint while a second, more methodical survey can then identify hot spots and areas in which responders should enter only when necessary and only with proper radiation survey instruments.

The initial survey is adequate to establish Hot Zone boundaries and to begin determining the initial public health measures that need to be taken outside of the Hot Zone, but it lacks the detail needed to work safely *inside* the Hot Zone. For that, it will be necessary to conduct radiological mapping inside the Hot Zone. This can be performed on foot using hand-held instruments, at ground level using remotely controlled vehicles (robots, "rovers," and the like) or by driving vehicles equipped with radiation detectors, from the air using helicopters or unmanned aerial vehicles

(UAVs, also called drones), or a combination of the above. The primary objectives of mapping the Hot Zone are to identify local hot spots in which radiation dose rates are high enough to potentially pose a risk to those working in those areas, to quantify general area radiation dose rates in the area, and to help determine low-dose ingress and egress paths for both responders and for members of the public when they are evacuated from the area.

It is important to recognize that local hot spots can represent more than a potential health risk; localized hot spots are places where radioactive sources, or pieces of radioactive sources might have landed or locations with high levels of contamination that might pose a resuspension and inhalation concern in windy conditions.

As these surveys are conducted the information must be recorded and transmitted to a central location so that it is available to the Incident Commander, public health officials, various affected agencies, and the elected and appointed officials with overall responsibility for responding to the event. There are several radiation instrumentation companies that have written software that will do this with data from their own instruments; some will incorporate information from any instruments that output data consistent with a standard protocol such as the American National Standards Institute (ANSI) Standard N42.42 (*Data format standard for radiation detectors used for Homeland Security*) [1]. In addition, various governmental agencies maintain software mapping systems that can be used to share information among all agencies at all levels of government that have been given access to this information.

15.2 Managing the Scene

There are two huge management challenges in the aftermath of a radiological or nuclear attack, managing the scene of the attack and managing the city as a whole. Managing the city as a whole in the aftermath of a nuclear attack is likely to be substantially different than managing the city in the aftermath of a radiological attack; for this reason managing the city as a whole will be deferred until Chap. 19 (Figs. 15.2 and 15.3).

Managing the scene of the attack will be the responsibility of the Incident Commander (under the American Incident Command System, or ICS) or the equivalent in other nations [13]. In the American system the Incident Commander will make use of their Command Staff, a team of subordinates, each with expertise in different areas such as Hazardous Materials, Health and Safety, Logistics, Investigations, and so forth. The ICS is designed to scale to the size and needs of the incident so that a small-scale incident might be managed without the need for a Command Staff, a simple fire might not require dealing with the risks posed by hazardous materials, and so forth.

A comprehensive discussion of managing the scene would require a book and, in fact, there are several book-length planning documents that have been developed by a number of nations and even several major cities. That being the case, a briefer discussion will have to suffice and we will touch upon matters of, health and safety,

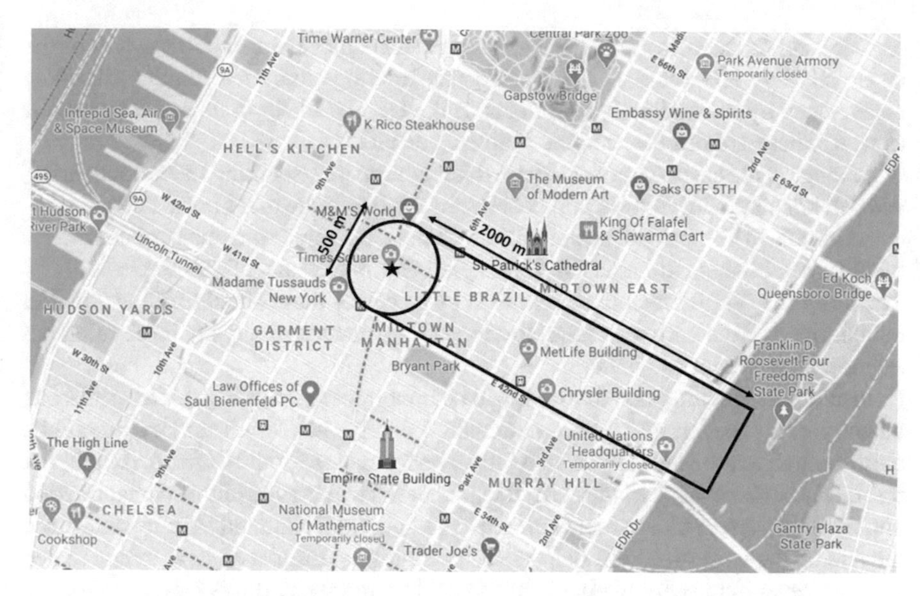

Fig. 15.2 Default Hot Zone following an RDD attack

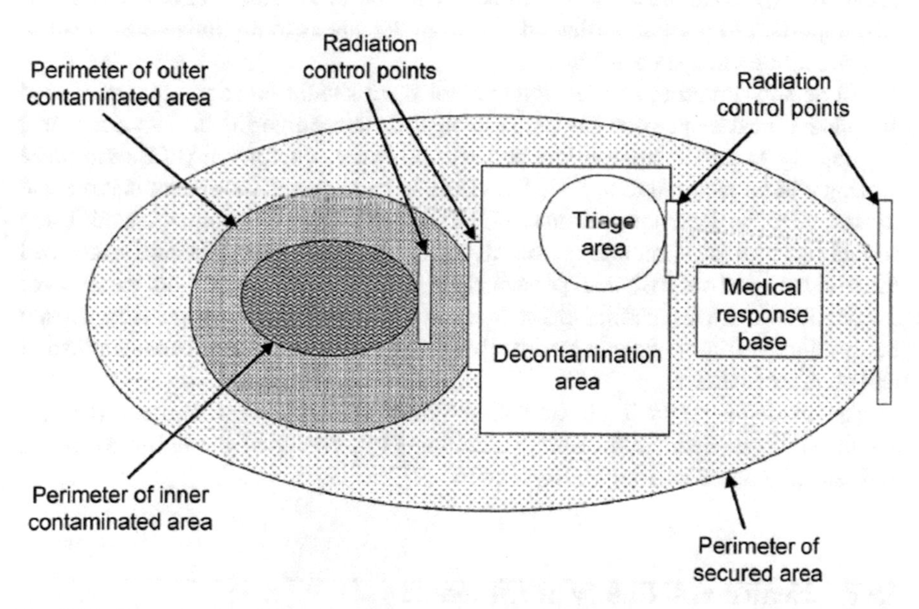

Fig. 15.3 Notional setup of Hot Zone and associated work zones. The Command Post would be in the vicinity of the Medical response base within the secured area. Equipment and personnel staging areas would be nearby. Used with permission of the National Council on Radiation Protection and Measurements, [7]

security and force protection, and investigating the attack. Not all of these will be the responsibility of the Incident Commander; the investigation, for example, will likely be overseen by the national government in the US and likely in many other nations as well. These are discussed in the following sections.

Managing the scene includes establishing boundaries, securing these boundaries, and establishing entry/egress points between the Hot Zone and the outside.

As noted in the previous section, the "First 100 min" document recommends initially establishing boundaries 250 meters from the site of the explosion and extending for 2 km downwind. As much as possible, these should follow "natural boundaries," so the outlines of the Hot Zone will almost certainly not be circular, but might well be easier to secure.

In addition to setting boundaries it will be important to establish a Command Post, an entry/exit corridor, and a decontamination point. The Command Post should be close enough to the scene to permit relatively easy access for the Command Staff, but it should also be in an area unaffected by the event with low levels of contamination and radiation levels close to background. Placing the Command Post upwind from the site of the explosion will help to ensure that it is not placed in a contaminated area; however, winds are likely to shift from time to time and it is inevitable that, at some points, the Command Post will be downwind of the site—for this reason it is also important to place it sufficiently far from the site as to minimize the spread of contamination via resuspension.

All persons entering and leaving the Hot Zone should do so via an established entry/exit corridor once one has been established. Those entering the Hot Zone must be properly garbed in appropriate personal protective equipment (PPE) and those leaving will be required to remove their PPE, to be surveyed for contamination, and to proceed to the decontamination station if they are found to be contaminated. There should also be a log sheet present that all personnel entering and leaving the area will use to sign in and out so that all personnel are accounted for. In addition, members of the public who are rescued should be noted as leaving the area as long as they are not badly injured—life-threatening medical concerns take priority over decontamination and all other matters.

The National Council on Radiation Protection and Measurements (NCRP) discusses the procedures for donning/doffing PPE, operating the entry/exit point, and related matters in detail in Report #161 [7].

15.3 Health and Safety at the Scene

Any emergency response is an inherently dangerous undertaking; emergency responders knowingly put themselves at risk entering dangerous areas in order to save members of the public and to try to limit the damage that is done. This might involve entering burning buildings, working in areas with toxic gases or ruptured utility lines, unstable structures, high radiation levels, working in the midst of heavy machinery, and more. In the second volume of NCRP Report 161 (*Management of*

Persons Contaminated with Radioactivity) [8] all of these risks, radiological and non-radiological, are summarized to help readers understand that, not only is radiation not the only risk present, but it is often not the most pressing risk facing emergency responders. Responders, the Health and Safety Officer, and the Incident Commander must all avoid the temptation to focus on radiological concerns to the exclusion of other factors that might pose a greater risk to their health and safety. The non-radiological hazards are fairly well-understood and are discussed in detail in many other documents (e.g. [4, 5, 14]); here we will concentrate on the radiological risks to emergency responders and to the public.

Among the highest priorities will be initiating lifesaving and rescue operations. It is essential to locate and evacuate those members of the public who are seriously or critically injured so that they can receive the medical care they need in order to survive. This means that, unlike emergency responders exiting the Hot Zone, members of the public who are badly injured can (and should) be evacuated to the hospital without delay, even if that precludes decontamination.

15.3.1 Internal Radioactivity, Inhalation and Ingestion

Inhalation and ingestion risks to emergency responders can be relatively easily managed by the use of respiratory protection and contamination controls. Respiratory protection will prevent inhalation of airborne radioactivity and will keep it from entering the mouth to be swallowed, while contamination controls will minimize the risk of accidental ingestion via the secondary contamination of food or drink.

However, inhalation can be a concern for members of the public who are in the area near where an explosive RDD was detonated and in areas downwind where the plume settles to the ground. Members of the public who are closest to the location of a bomb's detonation will be most at risk of inhaling radioactivity and will also be at the greatest risk of suffering from embedded radioactive particles and fragments from a source shattered by the explosion. In addition, those sufficiently close to the explosion will have injuries (e.g. cuts and scrapes) that provide a pathway for radioactivity to enter the bloodstream, as well as embedded fragments of radioactive materials.

From a public health perspective, one should assume that emergency responders, if properly equipped with protective equipment, are unlikely to have a significant intake of radioactivity. Thus, public health officials can devote the majority of their attention towards the goal of attempting to determine the intake of radioactive materials by members of the general public. This dose assessment process will help to determine the number of people requiring decorporation therapy to remove radioactivity from their bodies [9].

Assessing and quantifying this internal radioactivity and the resultant dose will be discussed in greater detail in Chap. 19.

15.3.2 Radiation Dose Management

If high levels of radioactivity are involved in the attack and radiation dose rates are high then managing radiation dose to the emergency responders will become a significant task. Part of dose management involves insuring that no individual exceeds a radiation dose limit, that the total dose of all workers combined is as low as reasonably achievable (ALARA), and that dose is balanced as much as possible rather than "burning out" a few people while others receive little or no dose.

The best way to accomplish this is by using dosimetry—ideally, every worker who enters the Hot Zone will have their own dosimeter, but this is not always possible. If there are insufficient dosimeters such that not every person can be issued one, it might be necessary to use group dosimetry, to be able to account for the location(s), time in each location, and dose rate at that location for each worker, or to use other dose estimation methods [10]. If possible, the dosimeters should be accredited by the National Voluntary Laboratory Accreditation (NVLAP) program, although if NVLAP-accredited dosimeters are not available, any type of dosimeter will suffice, including PRDs that record the total dose in addition to the dose rate.

Individual dosimetry involves issuing a dosimeter to each person entering the Hot Zone, and reading the dosimeter when that person exits. If there are insufficient NVLAP-accredited dosimeters to provide one for each person it is also acceptable to issue electronic dosimeters or PRDs that have a total dose feature that can be used to record the person's dose while in the Hot Zone. Either way, the most accurate reading will occur if the dosimeter is read or is zeroed out before the wearer enters the Hot Zone. If this is not possible, the wearer should read (and record) the total dose prior to entry and the difference between these readings represents their dose during that particular entry.

If there are insufficient dosimeters or instruments to provide to all persons entering the Hot Zone then group dosimetry should be considered. Group dosimetry involves providing an individual dosimeter for at least one person in each working group (e.g. firefighting hose team, rescue team, etc.) entering the Hot Zone, and the dose to that one individual is assigned to all members of their team. For group dosimetry to be relatively accurate the members of the group must be working in relatively close proximity in an area in which radiation dose rates are relatively constant from place to place (i.e. no person is likely to be standing in a localized hot spot).

In the absence of sufficient dosimeters for group dosimetry, it might be possible to calculate radiation exposure if the location of a person (or group) can be determined, as well as the amount of time spent in each location. A person (for example) who spent ten minutes in an area with radiation levels of about 60 mGy hr^{-1} should receive a dose of about 10 mGy from their work in that area. If they spend an additional six minutes in an area with a dose rate of 200 mGy hr^{-1} then an additional 20 mGy can be added to their dose for a total of 30 mGy for that entry. Assigning a radiation dose based on such calculations is possible, but this is not the preferred method for determining exposure.

In the initial moments after an RDD explodes it might not be possible to try to issue dosimeters or radiation detectors to those entering the area, and waiting for dosimeters to arrive might delay rescue operations, costing lives among the critically injured. In such cases it makes sense to send rescuers into save the lives of the badly injured, rather than to let people suffer or die needlessly. If this happens, the Incident Commander and/or Health and Safety Officer should keep track of each person who enters during this phase of operations and the length of time they were inside the Hot Zone so that a rudimentary dose reconstruction can be performed later. If possible, video recordings can also be used to help to constrain the work locations and the amount of time that passed during each entry. If staffing permits, the Incident Commander should also consider removing from radiological work all of those who made entries without wearing dosimetry until their dose can be estimated; these people can be moved to other duties such as driving, logistics, air sampling, and similar necessary duties outside of the Hot Zone.

15.4 Force Protection

It is not uncommon for terrorist groups to launch secondary attacks against emergency responders. These attacks can include setting explosives or placing snipers in areas in which responders are expected to congregate or placing teams tasked with further mayhem near the area attacked. Responding to an explosion is challenging; responding to a radiological attack is even more so. Adding secondary explosions and gunfire to the mix can make things even more difficult, especially if the primary attack took place in a crowded location and the secondary attacks panic the public. Not only that, but secondary attacks can deplete the number of emergency responders capable of assisting at the scene as well as disrupting all emergency response activities. For this reason, force protection is an important aspect of response to any terrorist attack.

Every military unit and law enforcement agency will have its own force protection plan and it is not the intent here to recommend details of force protection plans; rather, this will serve as a brief overview of some of the concerns that must be addressed by those specializing in force protection.

An explosion in a restricted location with limited exits will cause members of the public to rush towards those exits; these same exits also serve as choke points for arriving emergency responders. These exits, then, are logical places for a secondary attack to take place. A terrorist group can set secondary explosives and/or suicide bombers at these locations, to be triggered remotely at an appropriate time, it can station snipers and/or gunmen to target these locations, it can block these locations to trap people where they are relatively easy targets, or a combination of these.

Multiple teams were involved in the 2008 attacks in Mumbai India; between November 26 and 29 teams of terrorists carried out a dozen attacks against multiple targets using guns, explosives, and setting fires [2]. But the concept of conducting multiple simultaneous attacks pre-dates 2008; al Qaeda conducted simultaneous attacks against American embassies in Tanzania and Kenya, as well as its attacks against both World Trade Center buildings, the Pentagon, and the thwarted attack against (presumably) the White House on September 11, 2001. Other terrorist attacks around the world have also included secondary devices set with the purpose of injuring of killing emergency responders and/or members of the public in the aftermath of an initial attack as described above.

For this reason, many law enforcement and military groups will quickly move to conduct careful searches of areas in the vicinity of an attack as well as somewhat further afield to search for secondary explosives as well as jamming cell phone frequencies to try to block that method of activating explosive devices. These sweeps can be performed by any personnel who have been trained to recognize explosive devices, but anything that is found should be investigated by trained bomb technicians to determine what has been found and (if necessary) to remove or to render the device(s) safe. Areas checked should include chokepoints into and out of the area attacked, areas where the public congregates, and likely emergency responder command and staging areas.

Snipers and gunmen pose an additional threat to emergency responders and the public, so another aspect of force protection should involve searching for and neutralizing this threat. Such efforts can include establishing counter-sniper positions on rooftops and other elevated locations in the vicinity of the scene of the attack, the Command Post, chokepoints, and other areas where people congregate; they can also include actively searching (visually, with metal detectors, and/or via thermal imaging) the crowd for persons with weapons; thermal imaging, explosive trace detection, and dogs can be used to search for persons with explosives at the same time. Helicopters (if available) can also be used to search rooftops, bridge towers, and other elevated locations for adversary snipers.

If responders are free of worry about snipers and secondary explosives then they can give their undivided attention to the response efforts. In addition, they might be able to remove ballistic (bullet-proof) vests, which can be heavy, hot, bulky, and can hinder movement and response efforts. Thus, force protection not only helps to safeguard the public and emergency responders, but can help the responders themselves to work more effectively.

15.5 Radiation Safety at the Crime Scene(s)

The area where an explosion occurs is a crime scene—in the "First 100 min" document DHS recommends that the Incident Commander designate the area within 20 m of the detonation point to be a part of the crime scene. But the scene of the attack is likely not the *only* crime scene—any location where radioactive materials were stolen, stored, or made into the weapon; any location in which explosives were fabricated or stockpiled; any locations where waste from the bomb lab was disposed…these are all locations that bear investigation and that might become additional crime scenes. All of these locations must be investigated, and investigators must take care to remain safe from both radiological and non-radiological concerns. The manner in which investigatory work should be performed is beyond the scope of this book; here we will discuss how to conduct the investigation while remaining safe from radiation and any radioactive contamination that might be present. Please note that the next chapter contains a more general and comprehensive discussion of radiation safety work practices; what is discussed here is targeted towards working at crime scenes that have elevated levels of radiation and/or contamination.

For the sake of discussion we will assume that that the area(s) being investigated are thought to be associated with a group of suspected terrorists who are thought to be involved in an RDD plot; whether this plot has recently been interrupted, is in progress, or was recently successful changes little with the exception that the level of radioactivity (and the potential risk) will be higher if the main source is present. If an attack has been interrupted or has already taken place the amount of radioactivity in storage facilities, apartments, dumpsters, and related locations might be relatively low unless the workers were sloppy or careless in their work habits.

The agency(s) investigating the potential crime scenes will depend on the resources available and the policies of the national government and other jurisdictions involved. In some nations, any such event would be considered an attack against the entire nation with a national organization having jurisdiction over the crime scene(s) and assuming the role as the primary investigatory body—in the United States this would be the Federal Bureau of Investigation (FBI). In other nations, the local or regional law enforcement authorities might take the lead. And, even in the US, it might be necessary for the FBI to request assistance from local law enforcement agencies in the early stages of an investigation, before additional resources arrive, or to help secure multiple crime scenes.

15.5.1 Clearing the Room

Ideally, bomb technicians will have entered the room to clear it of explosives and potential hazards prior to the arrival of the radiological and hazardous materials team. However, the bomb technicians should have the ability to monitor radiation dose rates and their total dose while they are in the room. Ultimately it will be up to each department to determine whether or not their bomb technicians should be equipped with radiation detectors, dosimeters, or both so that they can monitor radiation levels as they work; or if they should be accompanied by a companion qualified to provide radiological coverage while the room is being cleared of explosives.

15.5.2 Prior to Entry

While approaching each location, radiation detectors should be turned on and monitored and, ideally, each person will have a dosimeter of some sort as discussed in

Sect. 15.3. If radiation dose rates are noted to be increasing as the investigators approach the location(s) of interest they should notify their chain of command in accordance with departmental policies and procedures. If dose rates on the approach (e.g. in the hallway leading to an apartment) reach a designated level (e.g. 20 μGy hr^{-1} the approach should stop at that point and an entry/egress corridor should be established at that location; otherwise the entry/exit corridor should be established sufficiently close to the area to be investigated as to permit relatively easy access to the area(s), but sufficiently far so as to provide a degree of safety from any materials—including explosives—that might be in the room. In addition, this area—where workers are donning and doffing their protective equipment, meeting to discuss survey results, planning the next entry, and so forth—should be located in an area where radiation dose rates are low enough to permit personnel to remain for a significant amount of time without approaching a dose limit. In the US, the default is to try to find a location with a dose rate of no more than 20 μGy hr^{-1}, although it might be necessary to accept a higher dose rate in order to maintain easy access to the work area.

All persons on the entry team should be briefed prior to entering the area(s); this briefing should include:

- The reason for investigating this particular location (e.g. human intel, elevated radiation dose rates, strange odors, etc.)
- Observations made during any earlier entries (if any)
- Any known or suspected risks present (e.g. chemicals, explosives, weak floorboards, etc.)
- Summary of any previous entries (e.g. bomb technicians) and their results
- Photos, drawings, sketches, etc.
- Any other relevant information about the scene
- Medical contingencies (e.g. location of first-aid kit, identity of any team members with CPR, first aid, or other medical training, location of nearest hospital emergency room or trauma center, etc.)
- Goal of entry (e.g. initial radiological scoping survey)
- Each team member's task(s) during entry (e.g. radiation survey, recording information, managing entry/exit corridor, radiological survey of persons leaving area, etc.

Following the briefing the entry team should don their protective equipment, reset the accumulated radiation dose on their electronic dosimeters to zero, and enter the area via the entry/exit corridor. The other personnel at the scene should remain in areas in which radiation dose rates are less than 20 mGy hr^{-1} unless their presence is required to perform decontamination, bring tools or other equipment to the entry team, assist with decontamination, or other duties. When all team members understand their duties and are ready to do so the entry team can perform the entry. The entry of each person should be logged to permit accounting for each person upon egress.

15.5.3 Initial Entry

The team performing the initial entry should consist of the smallest number of people who can perform the required tasks. While a single person might be able do so, for safety reasons a two-person team is preferred when possible unless the room is small (e.g. a shipping container, storage area, or room of comparable size). For a two-person team, each person's duties might include:

Data collector monitors radiation dose rates at each of the five survey locations (from the "Z-pattern survey" described below) and notes other risks and significant safety information (e.g. the presence of chemical containers, spilled powders or liquids, injured or deceased persons, signs of fire, explosion, corrosion, locations of boxes, backpacks, suitcases, and other potential hiding places, evidence of possible biological agents, and any other information regarding potential safety concerns during the investigation.

Recorder writes down each observation, including the radiation dose rates at each survey location and the location of all potential risks and objects of interest noted by the data collector.

Upon completion of the initial radiation scoping survey the entry team will exit the area through the entry/exit corridor, removing their PPE and surveying themselves for contamination and decontaminating as necessary. After the entry team has cleared the survey and decontamination station the entire team will review the radiological data and other observations that were noted and will determine the course of their subsequent investigation, including both a more detailed radiological survey and appropriate methods for neutralizing, removing, or working around the hazards noted.

The radiation safety practices followed during the initial entry are fairly straightforward and are similar to those described in the following chapter for working in the RDD Hot Zone. The primary difference is that, at a location where terrorists might have been producing explosives for a radiological weapon—and possibly other WMD as well—all materials, containers, bags, furniture, cupboards, drawers, and so forth must be assumed to contain dangerous materials and to be booby-trapped in some manner until proven otherwise (Fig. 15.4).

In this graphic, the person conducting data collection enters the room and travels quickly to each of five survey points—all four corners of the room and the center. At each location they report the radiation dose rate measured and then proceed quickly to the next location, noting other observations (e.g. presence of chemical spills and containers). In the example shown here, the readings suggest that a radioactive source is located along the back wall of the room and a second entry would be performed to determine the precise location. Once located and photographed, the source should be removed or shielded so that another survey can be conducted to look for the presence of additional sources.

1. Chemical containers on floor
2. Spilled powder
3. Metal can lying on side with no lid
4. Waste can filled with PPE
5. Metal shavings on workbench
6. Papers with radiation symbol
7. Body
8. Lead pig

Location	Reading
1	100
2	200
3	300
4	200
5	900

Fig. 15.4 Initial scoping survey, including both radiation dose rates and other safety-related observations

15.5.4 Radiation Safety During the Investigation

Once the initial entry has been completed and the investigation has begun, radiation safety practices should settle into a routine as will be described in the following chapter, with a few additions. Note, too, that in a non-emergency situation there is no need for those entering the radiological areas to violate routine dose limits for radiation workers or for non-radiation workers (as appropriate).

- Follow-up radiation and contamination surveys
  - The initial entry will concentrate on gamma radiation dose rate because gamma radiation can be detected at long range and it can pose an immediate danger to life and health.
  - However, other types of radiation might be present and surveys during subsequent entries should include surveys for alpha, beta, and neutron radiation as well as identifying the gamma-emitting radionuclide(s) present.
- Evidence collection
  - Radioactive contamination itself is evidence as it likely originated from one or more sources that were used (or were to be used) in an attack. As such,

items and areas should not be decontaminated until the contamination has been properly characterized.
- In addition, evidence found at the site should not be decontaminated until all forensic evidence (e.g. contamination distribution patterns, nuclide(s) present, etc.) has been noted and recorded.
- Evidence that is bagged for removal from the site should be double-bagged and labeled as contaminated, preferably using the radiation trefoil, to alert workers at the forensics laboratory to take appropriate precautions.
- Otherwise, all normal evidence collection, documentation, and transportation protocols should be performed in accordance with normal departmental policies.
- Reduce exposure from high-activity sources by placing them into shields, placing portable shielding around them, or removing them from the area(s) being surveyed after photographing them to record their location, appearance, and condition as they were found.

- Dose management
  - If radiation dose rates in the Hot Zone are in excess of 250 μGy hr^{-1} the person in charge at the scene should consider assigning stay times to all personnel entering the Hot Zone.
  - All persons issued self-reading (active) dosimeters should be reminded to monitor them periodically and to report their readings to the person in charge at the scene and/or the Health and Safety Officer.

As noted above, all personnel at the scene need to keep the radiological risks in perspective, given the potential threats posed by hazardous chemicals, explosives, toxic fumes and vapors, flammable substances, and more. Those working at these locations cannot pay so much attention to radiological issues that they fall prey to these other, often more substantial hazards.

References

1. American National Standards Institute (2013) Data format standard for radiation detectors used for homeland security (Standard N42.42). Institute of Electrical and Electronics Engineers, New Jersey
2. Government of India (2009) Mumbai terrorist attacks (November 26–29, 2008), Dossier presented to the Government of Pakistan. 2009. https://fas.org/irp/eprint/mumbai.pdf. Accessed 19 Nov 2020
3. Harper F, Musolino S, Wente W (2007) Realistic radiological radiological dispersal device hazard boundaries and ramifications for early consequence management decisions. Health Phys 93(1):1–16
4. Jackson B, Baker J, Ridgely M, Bartis J, Linn H (2004) Protecting emergency responders, vol 3. Safety Management in Disaster and Terrorism Response. RAND, Santa Monica CA

5. LaTourrette T, Peterson D, Bartis J, Jackson B, Houser A (2003) Protecting emergency responders, vol 2. Community Views of Safety and Health Risks and Personal Protection Needs. RAND, Santa Monica CA
6. Musolino S, Harper F (2006) Emergency response guidance for the first 48 hours after the outdoor detonation of an explosive radiological dispersal device. Health Phys 90(4):377–385
7. National Council on Radiation Protection and Measurements (2008) Report No. 161, Management of Persons Contaminated with Radioactivity, vol 1. NCRP, Bethesda MD
8. National Council on Radiation Protection and Measurements (2008) Report No. 161, Management of Persons Contaminated with Radioactivity, vol 2. NCRP, Bethesda MD
9. National Council on Radiation Protection and Measurements (2010) Report No 166: population monitoring and radionuclide decorporation following a radiological or nuclear incident. NCRP, Bethesda MD
10. National Council on Radiation Protection and Measurements (2017) Report No 179: guidance for emergency response dosimetry. NCRP, Bethesda MD
11. Stallings R (1984) Evacuation behavior at three mile Island. Int J Mass Emerg Disasters 2(1):11–26
12. US Department of Homeland Security Science and Technology Directorate (2017) Radiological dispersal device (RDD) response guidance: planning for the first 100 minutes. US DHS, New York City
13. US Federal Emergency Management Agency (2017) National Incident Management System, 3rd ed. Washington DC
14. Willis H, Castle N, Sloss E, Bartis J (2006) Protecting emergency responders, vol 4. Personal Protective Equipment Guidelines for Structural Collapse Events. RAND, Santa Monica CA

Chapter 16
Working Safely in a Radiological Area

While we saw in Chap. 2 that radiation is not as dangerous as most believe, it can still pose a risk of both short-term (deterministic) and long-term (stochastic) health effects if workers do not take adequate precautions. It is important that those working in any radiological area be able to evaluate the risks that are present and that they take precautions to minimize their risks and subsequent health effects. Report #161 of the National Council on Radiation Protection and Measurements [4] offers a thorough summary of radiation safety practices in response to a number of incidents that might result in the contamination of personnel. Rather than simply repeat what is in this (and other) reports, it seems more appropriate to summarize the most relevant points and to add additional considerations that are relevant to working on the site of a radiological and nuclear interdiction and attack; radiation safety precautions that are specific to the aftermath of a nuclear attack will be discussed in Chap. 18.

16.1 General Radiation Safety Good Work Practices

There are some radiation safety work practices that are common to most situations, including both routine and emergency situations. Briefly, these are [5]:

- **Time**—minimize the amount of time spent in the Hot Zone in general, and in areas with high dose rates in particular. This can be accomplished through
  - Pre-planning and briefing each entry so that each person understands the objectives, their specific tasks, the location of hot spots, and other relevant information about the entry.
  - Practicing complex tasks prior to entry if possible.
  - Working quickly, but not hastily.

P. A. Karam, *Radiological and Nuclear Terrorism*,
Advanced Sciences and Technologies for Security Applications,
https://doi.org/10.1007/978-3-030-69162-2_16

- **Distance**—maintain the greatest distance possible from any identified radioactive sources and areas with the highest radiation dose rates; even relatively small changes in distance (e.g. working at arm's length instead of standing directly adjacent to a source) can reduce exposure considerably.
- **Shielding**—workers can install temporary shielding, which can be lead sheets, soil, or even large water bladders such as those used as water supplies by fire departments. Workers can also make use of existing structures (e.g. masonry walls), vehicles, buildings, and terrain to reduce radiation exposure.
- **PPE**—wearing PPE can reduce skin dose by keeping contamination from direct contact with the skin and can reduce internal exposure by protecting the mouth and nose using a respirator. Covering open cuts, scrapes, and other injuries can also help to reduce internal exposure.
- **Dosimetry and radiation instruments**—since radiation cannot be sensed, the only way we have to know if it is present and in what levels is through our radiation instruments. Responders should not enter a radiological area without dosimetry and/or radiation detectors unless victims are going to die without immediate attention (Fig. 16.1).

Fig. 16.1 Two workers searching for a radioactive source at a municipal waste transfer station. Note the radiation detector in the hand of the person on the left and the PPE (including respiratory protection) on both workers. Author's photo

16.2 Radiation Safety Considerations Relevant to a Radiological "Dirty Bomb" Interdiction and Response

While there are a number of radiation safety practices that are common to a large number, if not all, radiological circumstances, a radiological attack raises additional issues. This is due, in part, to the fact that an accident involving the incidental release of radioactivity is different than the malicious release of radioactivity with the intent to cause harm. Consider: in 1987 an accident involving a 50.9 TBq Cs-137 source resulted in four deaths, 120 individuals who required medical care, and 250 who required decontamination, and the cleanup created several thousand cubic meters of radioactive waste [2]. Forty-four TBq, approximately 86% of the original activity, of Cs-137 was recovered and accounted for. By comparison, a terrorist attack would likely involve the malicious release of nearly 100% of the radioactive material. Compared to a radiological accident, a terrorist attack is likely to involve:

- The release of a greater percentage of radioactivity from the source(s) used
- A larger contaminated area with contamination spreading further downwind
- Large amounts of airborne radioactivity during the initial release
- A wide variety of particle sizes due to heat and pressure from the blast
- Larger numbers of people with internal contamination due to absorption into wounds, embedded fragments, and inhalation of airborne radioactivity
- Possible high-activity sources or source fragments scattered within a few hundred meters of the site of the explosion
- The presence of damage to roads, structures, utilities, etc. complicating response efforts
- Infrastructure contamination, including streets, sidewalks, sewers, etc.

Some of these do not affect radiation safety much, if at all. Most, however, require that responders in the Hot Zone **and any other areas affected by the attack** be aware of the potential concerns and take appropriate protective actions to avoid harm. Workers at waste water treatment plants and those entering sewers downstream of the area attacked, for example, will need to be taught to use radiation instruments and to work safely in radiological areas until the sewers and treatment plants can be surveyed and characterized.

Unlike an accident, a radiological attack is likely to result in the release of a greater percentage of radioactive material from the source(s) used than would occur during an accident. This means that radiation dose rates will be higher, more people might be seriously affected, and the contamination is likely to affect a larger area than would be the case in an accident. Thus, the Hot Zone is likely to be larger and concomitantly more difficult to secure than was the case in Goiânia and other accidents. In addition, responders must be aware of their accumulated radiation exposure as they work to avoid exceeding a dose limit. It is also possible that the heat and pressure of the explosion can cause physical and chemical changes to the radioactive materials being dispersed; this can cause powders to clump together (for example) into larger particles that will behave differently when passing through the air than will powders.

The physical conditions of a blast can also cause solid sources (e.g. metal alloys or ceramics) to fragment and spall, creating a large number of higher-activity bits of radioactivity scattered through the area affected by the blast. These fragments can create localized hot spots within the Hot Zone, can be embedded in the flesh of persons within a few hundred meters of the explosion, and can pose a risk to responders who might encounter them or try to pick them up, potentially leading to skin burns and other tissue damage.

The explosive dissemination of radioactive materials, especially if that material is in a physical form that is easily dispersible (e.g. liquid or powder), is likely to blast the radioactive material to altitudes of a few hundred meters or higher, putting the dust and debris into higher-level winds which will carry it further than would be the case at lower levels. This also increases the amount of radioactivity available to be inhaled over a large area, increasing the number of people exposed to airborne radioactivity. This wider spread of radioactivity can also cause the contamination of infrastructure close to the site of the explosion as well as areas further afield, including sewer systems as firefighting water flushes the contamination away from the scene and into storm sewers and the city's waste water treatment system.

The explosive dispersion of radioactivity can also lead to injuries (cuts, scrapes) through which radioactivity can be absorbed into the bloodstream as well as radioactive particles that can be driven into the body by the force of the explosion. This increases the possibility of a radioactive intake with the resulting internal contamination and exposure to those close enough to the blast to be affected.

On top of these radiological effects, the explosion of an RDD can cause physical damage to proximate areas—the damage to utility lines, roads and bridges, and the contamination of additional areas can hamper the movement of responders and their equipment into the affected area as well as the movement of injured persons from the area to receive medical care. While these are not, in and of themselves, radiation safety concerns, they can cause increased radiation exposure to persons who are contaminated or affected by embedded fragments of radioactive materials.

16.3 Other Safety Considerations Relevant to a Radiological "Dirty Bomb"

Any explosion creates risks, and larger explosions create more and greater risks, and focusing on the radiological aspects of an attack can take one's attention away from the non-radiological threats. While the NCRP [4] listed a number of these in Report #161, most that are relevant here were discussed in Chaps. 6 and 7 of this book. However, there are additional concerns that are relevant to emergency responders that either were not raised by the NCRP or that have not been discussed here—or both.

- *Secondary devices* and/or unexploded primary devices might be encountered in the area that was attacked and elsewhere in the city. These might be obvious (e.g. a pipe with wires and batteries stuck to it) or they might be disguised or hidden. For this reason, the bomb squad should clear all areas, including those within the Hot Zone (with appropriate radiological support) as soon as possible. Until cleared by bomb squad personnel, all containers, packages, dumpsters, trash cans, and even piles of trash should be assumed to contain explosive devices that can be set off remotely, via trip wires or other triggering devices, or through physical contact (e.g. bumping, kicking, picking up, etc.).
- *Unstable structure and streets*, can pose a threat to responders working in the area. An explosive device, for example, might weaken streets and sidewalks and in a city with an extensive underground infrastructure, this can lead to the pavement collapsing into utility or train tunnels, endangering anybody entering these areas. The blast can also weaken structures, making them dangerous to enter and dangerous to work around. In addition, while members of the public should seek shelter in the moments following any explosion, if they seek shelter in an unsafe or unstable building, they will have to be told to evacuate to a safer building nearby. This means that emergency responders might have to enter an unstable structure in order to evacuate those inside to a safer location. Such entries must be made carefully and deliberately, in accordance with whatever departmental procedures and policies cover such work.
- *Ruptured utility lines and other structures* can pose significant risks and/or can hamper response efforts:

  - Electrocution (electrical lines)
  - Fire or explosion (natural gas lines)
  - Flooding, drowning, erosion and undermining pavement (water mains)
  - Infection and disease (sewer)
  - Loss of communications (telephone, cable, fiber-optic lines)
  - Collapse of pavement and tunnels.

- *Emergency response activities* can also pose a risk to those working on the response. Trucks, ambulances, and other vehicles pose a risk to emergency responders working in the area, as do other emergency response activities. People can be injured or killed by the response efforts themselves; for this reason, it is essential that everybody working in or transiting through the work zone remain aware of their surroundings at all times. In particular, when it is time to evacuate members of the public from the Hot Zone the Incident Commander must take into account the work taking place; it might be advisable to stop work and vehicle movement during the evacuation to minimize risks to the public and to the responders who are guiding them from the Hot Zone.

16.4 Radiological Health and Safety Officer

Most agencies involved in responding to emergencies will appoint a Health and Safety Officer (HSO) to see to the welfare of the agency's personnel at the site. Depending on the city's Incident Command System and the scale of the event there might be a single HSO for the entire site or each agency might have their own HSO. A person in this position can directly or indirectly affect the health of every person responding to the incident, making it important that, if possible, the HSO have appropriate training and experience to carry out their duties competently [3]. The duties of the HSO at the scene of a radiological incident include [1]:

- Radiation dose management
  - Dosimetry and instrument issue
  - Remind personnel in Hot Zone to read electronic dosimeters periodically
  - Assign stay times and/or dose limits to personnel entering Hot Zone
- Hot Zone management
  - Determine and administer PPE requirements for work in Hot Zone
  - Maintain Hot Zone entry/exit logs
- Environmental sampling
  - Conduct or order radiation surveys, air samples, contamination surveys, and so forth as appropriate
  - Review environmental sampling results and adjust entry requirements as appropriate
- Record-keeping
  - Ensure appropriate records are maintained as called for by policies and procedures
  - Collect records at end of each shift and hold for later reference and use (e.g. to help adjudicate future compensation claims, for individual occupational health records, etc.).

In addition to the radiological concerns, the HSO also needs to keep track of the other health and safety issues as well—heat stress, hydration, breathing apparatus constraints (e.g. stay time for air supply), and more; it might make sense for some of these duties to be delegated to subordinates or to delegate the radiological health and safety duties to someone with expertise in this area.

16.5 Radiation Safety Policies and Procedures

Every organization that anticipates being involved in radiological work should develop policies to describe the conditions under which their personnel can be exposed to radiation and radioactivity as well as the limitations that apply to this work. For example, a fire department might limit Hot Zone entries to only those firefighters with radiation safety and hazardous materials training or a police department might require that everybody investigating a radiological crime scene must make use of individual or group dosimetry while conducting an entry into a radiological area. Policies such as these—and many more—help to ensure that personnel working in these areas will work safely and in accordance with good radiation safety work practices. The author has been involved in developing procedure and policy manuals for both law enforcement and fire departments; while these manuals cannot be shared, the sections that comprise them can be and include subjects such as:

- Responsibility for administering radiation safety at the scene of an attack or investigation
- Maintaining radiation exposures As Low As Reasonably Achievable (ALARA)
- The duties and responsibilities of the Health and Safety Officer at a radiological scene
- Radiation dosimetry requirements
- Training requirements for radiation workers and for non-radiation workers at the scene of a radiological emergency
- Radiation dose limits under emergency and non-emergency conditions
- Use and maintenance of radiation instruments
- Proper use of radiation labels, signs, postings, etc.
- Performing and documenting radiological surveys
- Retention of records from radiological work
- Taking possession of lost or stolen radioactive materials
- Other subjects as applicable to the specific agency.

Each of these policies will give rise to a set of procedures helping the affected personnel to understand how to apply and comply with that policy in practice. For example, the radiation dosimetry policy might include procedures such as:

- How to issue a radiation dosimeter
- How to wear and use a radiation dosimeter
- Using a PRD in lieu of radiation dosimetry
- Reading radiation dosimetry
- Record-keeping
- Using group dosimetry.

Radiation worker training (and the annual refresher training) should include information about the policies and procedures that are relevant to their duties, and these should also be evaluated during drills and exercises.

References

1. Federal Emergency Management Agency (2017) Resource typing definition for the national qualification system emergency management: safety officer (FEMA 509-v20170717). FEMA, Washington
2. International Atomic Energy Agency (1988) The radiological accident in Goiânia. International Atomic Energy Agency, Vienna
3. Karam P (2020) Suggested training and experience qualifications for health and safety officers during a radiological incident. Health Phys 118(4):458–461
4. National Council on Radiation Protection and Measurements (2008) Report No. 161, Management of persons contaminated with radioactivity, vol 1. NCRP, Bethesda
5. Strom DJ (1996) Ten principles and ten commandments of radiation protection. Health Phys 70(3):388–393

Chapter 17
Medical Response

Any attack, radiological or nuclear, will cause harm to many people and these injured people will require medical attention. A number of governmental and other agencies have developed recommendations, guidelines, and model procedures for addressing the aftermath of radiological and nuclear attacks, including the International Atomic Energy Agency [16, 17], the State of Washington [27], and the National Council on Radiation Protection and Measurements [24]. In addition, a number of experts from academia and medicine have also contributed valuable insights, drawing from experiences over the course of decades (e.g. [8, 12]) that have included radiological and nuclear accidents as well as the nuclear bombings in Japan in 1945. The result is a fairly robust understanding of medical issues that will be faced in the aftermath of a radiological or nuclear attack. That being said, this does not mean that the medical effects of such an attack would be easily addressed because there are very few physicians who have actual personal experience in this area. Luckily, many of those who do have personal experience have written or contributed to the writing of books, book chapters, and papers on this subject.

The American College of Radiology [2] notes that, in the event of a radiological attack, medical facilities and staff will be required to carry out ten basic tasks:

1. Initially focus on treating injuries and stabilizing the patient in the event of an incident that combines radiation exposure and physical injuries
2. Hospitals must prepare to manage large numbers of frightened people and may be overwhelmed at first
3. Hospitals must have a plan to distinguish those requiring medical care from those who can be sent to an off-site facility (e.g. a Community Reception Center)
4. Hospital staff must know how to set up an area for treating radiological victims
5. Hospital staff should approach decontaminating a radioactively contaminated patient as though the patient is contaminated with raw sewage, taking similar precautions

P. A. Karam, *Radiological and Nuclear Terrorism*,
Advanced Sciences and Technologies for Security Applications,
https://doi.org/10.1007/978-3-030-69162-2_17

6. The hospital should minimize the spread of contamination by using double sheets and stretchers to transport patients from the ambulance to treatment areas
7. Staff should be able to recognize and treat patients who have been exposed to high levels of radiation
8. Staff should be able to recognize radiation injury
9. The hospital should maintain a list of agencies and organizations to contact in case of a radiological emergency
10. Hospital staff should develop a plan to evaluate and counsel uninjured patients who were exposed to radiation.

Some of these are fairly straightforward radiation safety problems; the difficulty is that emergency and medical responders are not radiation workers and what is relatively simple and straightforward to radiation workers is not necessarily known to emergency or medical responders. In addition, the relative lack of understanding as to the health effects of radiation can cause medical responders to be more concerned about the dangers of radiation than is warranted. This, in turn, can lead them to delay needed treatment or to give radiological concerns a higher priority than they ought to have.

Many of these issues will be discussed in this chapter.

17.1 Medical Care at the Scene

Medical care at the scene of an explosion should focus on largely the same things whether the explosion includes radioactivity or not; as has been noted elsewhere in this book, priority must be given to life-endangering medical issues (e.g. arterial bleeding, amputated limbs, crush injuries, etc.) followed by lesser injuries. If it is high, the radiation dose rate might complicate rescues at the scene, but will drop off quickly with distance from any source(s) and will have little effect on medical care provided outside of the Hot Zone. Contamination levels might also complicate the work that needs to be performed due to requiring personnel to wear appropriate PPE and possibly to perform decontamination, but it will rarely be a controlling factor (Fig. 17.1).

When performing rescues at the scene of an attack the primary consideration should be the safety of those being rescued and of the responders performing the rescue. If conditions at the scene are more or less safe and stable then the safest course of action might well be to stabilize the person being rescued before moving them from the area; this avoids causing or exacerbating injuries that might occur when moving the victims. If, however, conditions at the scene are potentially dangerous due to fire, unstable structures, ruptured or broken utility lines, high radiation dose rates, or other factor(s) then the victim(s) should be moved to a location that poses little risk in and of itself so that the victim can be stabilized in safety. In addition,

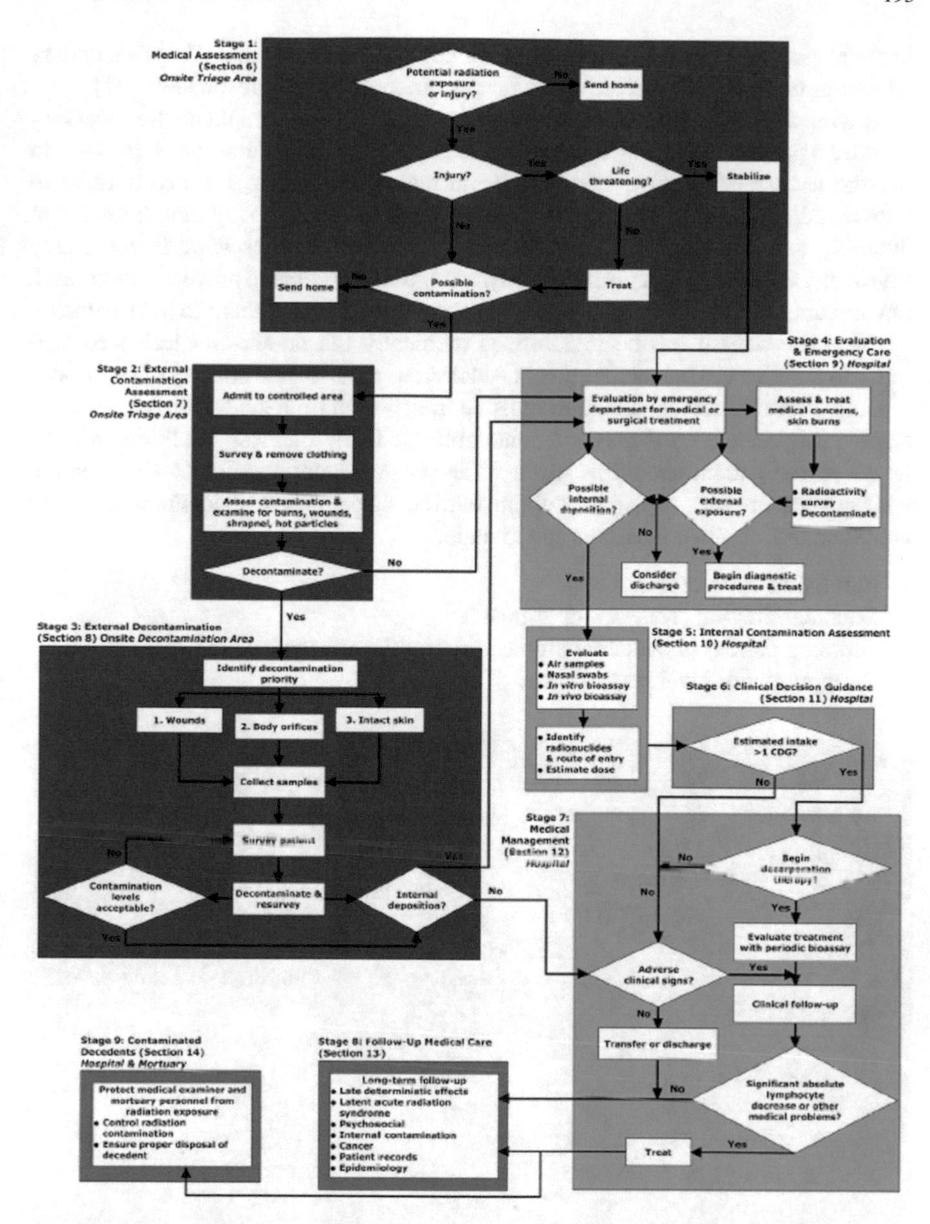

Fig. 17.1 Decision tree for management of persons contaminated with radionuclides. With permission of the National Council on Radiation Protection and Measurements [24]

medical responders at the scene face the same risks, including the risks of secondary attacks aimed at them, that are faced by any other emergency responders [33].

Having said that, responders should also understand that it might not be necessary to move a person all the way to the perimeter of the Hot Zone; in some cases it might only be necessary to move a few or several meters away from a source in order to reduce radiation dose rates to levels that will permit a longer stay time (remember, doubling one's distance from a source will reduce dose rates—and increase stay times—by a factor of four). Accordingly, it might be possible to move a person just a few meters, stabilize them in this location, and then evacuate them to the perimeter.

Once the patient has been stabilized medically the on-site medical personnel must do their best to balance medical needs versus radiological concerns. Those who urgently require medical attention must be transported immediately to a hospital to receive this care, even if they are contaminated. If the medical condition permits, however, the patient should be given whatever decontamination or contamination control measures are possible without putting them at risk. Decontamination or contamination control measures can include:

- Removing outer clothing
- Decontamination (wet, dry, or "moist")
- Dressing patient in contamination control outfit over their clothing
- Wrapping with blankets or sheets.

How far to reach an area with a safe radiation dose rate?

Arriving at the scene of a bombing about 30 min after an explosion was reported the initial entry identifies a badly injured person at a distance of about 1 m from a radioactive source. Your radiation detector alarms at a distance of about 30 m from the source, indicating a dose rate of 10 mGy h^{-1}. Using the inverse square law, dose rate at a distance of 1 m will be higher by a factor of 30^2, or about 9 Gy h^{-1}. At that dose rate, a person will receive a fatal dose of radiation in less than one hour.

Thirty minutes after the attack, victim has already received a dose of 4.5 Gy and taking 15 min to stabilize the victim in place is likely to expose them to a cumulative dose high enough to be potentially fatal, and you will receive a dose high enough to cause severe radiation sickness. At the same time, their injuries are potentially severe and you judge that they are unlikely to survive evacuation to the perimeter (100 m away) without being stabilized. How far do you need to move them so that they can be stabilized without putting them at risk from radiation exposure?

If your goal is to keep the total radiation exposure to less than 5 Gy then they need to be moved to a distance at which they will receive 0.5 Gy in 15 min, which is a dose rate of 2 Gy h^{-1}.

Reducing the dose rate from 9 to 2 Gy h^{-1} requires reducing the exposure rate by a factor of 4.5. Since dose rate drops with the square of distance, the distance must increase by a factor of $\sqrt{4.5}$, or by a factor of 2.1 —moving the victim to a distance of 2.1 m will reduce radiation dose rates sufficiently to stabilize the victim to make it safe to remove them the rest of the way to the perimeter.

Realizing that it is not easy to calculate square roots without a calculator under such circumstances, one can also simply realize that the dose rate at a distance of 2 m will be ¼ the dose rate at 1 m (about 2.25 Gy h^{-1}) and at 3 m the dose rate will be lower by a factor of 9 (1 Gy h^{-1}). Moving the victim 2 m further from the source will get them to a dose rate that will make it possible to safely stabilize them.

Emergency responders at the scene should use their judgement to apply as much contamination control as possible without affecting the person's medical condition. For example, a person who is gravely injured should be transported to the hospital without delay; a person with relatively minor injuries should be decontaminated prior to transportation to avoid needlessly contaminating an ambulance and emergency room.

In all operations in the Hot Zone emergency responders must be careful to keep their exposures low enough to keep from becoming ill themselves—less than 1 Sv—if at all possible. This means that individual responders need to monitor their dosimetry (or the total dose feature of their radiation instruments), and the Incident Commander and/or the Health and Safety Officer must remind those in the Hot Zone to check their dosimeters frequently and to exit when they reach a dose limit (actually, somewhat sooner as it will take time to exit the area).

The NCRP [24] and REAC/TS [28] have each developed very similar flowcharts for the triage, evaluation, stabilization, and radiological assessment of those injured during a radiological event. As noted elsewhere in this book, the primary consideration must be attending to urgent medical needs and stabilizing the patient; evaluating the person's exposures to internal and external radiation and contamination is a secondary consideration as even a fatal dose of radiation is unlikely to prove fatal in less than several days or weeks while medical concerns (e.g. arterial bleeding) can be fatal in minutes or hours.

Critically injured patients should be taken through the entry/exit corridor and directly to the hospital. All others should be brought out of the area and taken to a triage area for medical and radiological evaluation. Once out of the area, medical personnel at the scene (e.g. paramedics and emergency medical technicians) should provide what treatment they can, including collecting information that can be used for radiological dose assessment at the hospital or at a Community Reception Center (Fig. 17.2).

Among the tasks that should be performed if the patient's condition permits are performing and recording a whole-body radiological survey to determine levels of skin contamination and the presence of any embedded source fragments. Survey

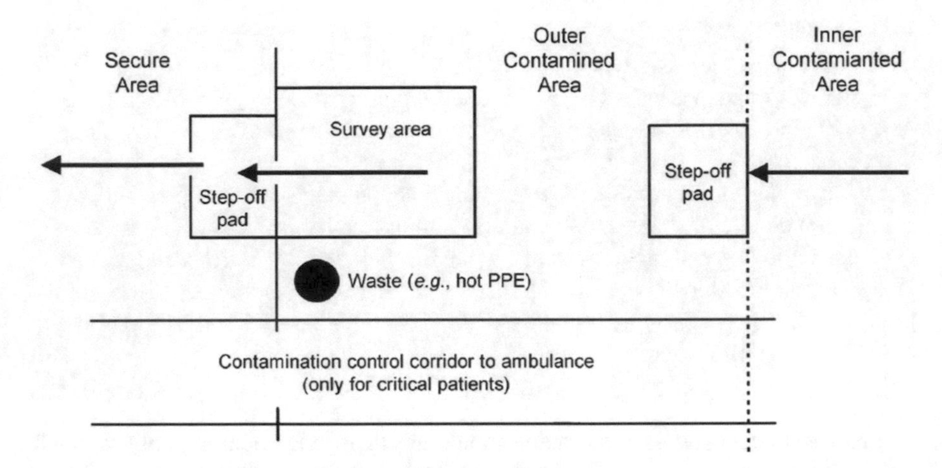

Fig. 17.2 An example entry/exit corridor for those entering and leaving the Hot Zone. Used with permission of the National Council on Radiation Protection and Measurements [24]

results should be entered onto a survey map that will be sent with the person to their next destination (hospital, decontamination area, home, etc.).

Views on obtaining nasal swabs are mixed because there are so many factors that can cause radioactivity to be lost, causing an apparently lower intake. Blowing the nose, wiping the nose, or having a runny nose, for example can remove radioactivity: breathing through the mouth can cause lower levels of contamination to be deposited in the nose, and so forth. The NCRP [24] concluded that, while the *presence* of contamination on nasal swabs can indicate a likely intake, the *absence* of contamination does not necessarily indicate that no inhalation took place. In either case, it can be useful to use nasal swabs as one of several factors that can be evaluated together to determine the likelihood of internal contamination, along with contamination on the head and face, elevated counts on the back or chest following decontamination, and so forth [4, 19]. If nasal swabs are to be obtained it is best to do so as soon as possible after a potential intake occurs.

Once decontaminated, those who require medical attention should be transported to a hospital or other center where they can receive the care they need, according to the city's mass casualty plan. Responders at the scene should also consider that a large number of people are likely to self-evacuate, often arriving at the hospital before the critically injured. In New York City, for example, the first to arrive at nearby hospitals following the attacks on September 11, 2001 were those who walked, drove, or took taxis. All relevant information (survey maps and results, decontamination conducted, medical information, and so forth) should be sent with each patient when possible (Fig. 17.3).

First Name: ______________ Middle Initial: ______ Last Name: ____________
Date of Birth: ____________________ Phone: ______________________
Address ______________________
Date/Time: ____________________ Drivers License # ______________
Location at time of incident: ______________________________
Parent or Guardian (if child): ______________________________

Mark contamination locations and survey reading on the diagrams below.

Circle if readings are in cpm mR/hr μR/hr

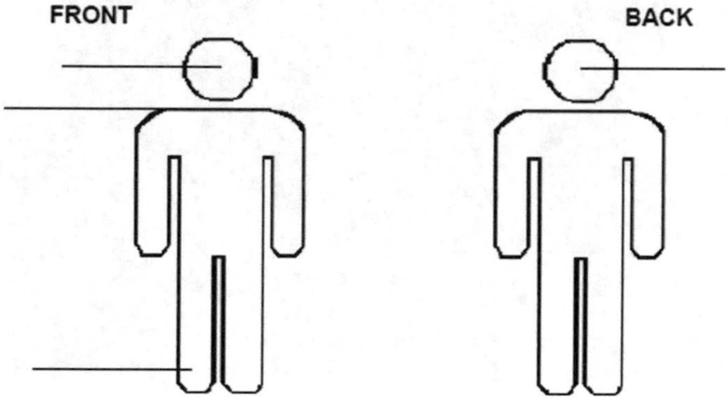

FRONT BACK

Survey results
<1,000 cpm__________ >1,000 cpm__________ >10,000 cpm__________

Comments: ______________________________

Monitored by: ______________________________

Person sent to decontamination area: _____Yes _______No
Clothing and valuable bag number: _____ Valuables returned: _____Yes _____No
Nasal area reading of 100,000 cpm or 0.5 mR/hr : ___Yes _____No
If Yes, refer to medical facility Person sent to medical facility: _____Yes ____No

Fig. 17.3 Example of a radiation survey form for recording surveys performed on individuals leaving the Hot Zone

17.2 Preparing the Hospital to Receive Contaminated Patients

Some of the preparation for receiving contaminated patients should come before there is an emergency; hospitals that anticipate caring for patients who are contaminated with radioactivity, chemicals, or other hazardous materials should establish at least one examination and treatment room suitable for receiving contaminated patients. In particular, any room designed for use with contaminated patients should include

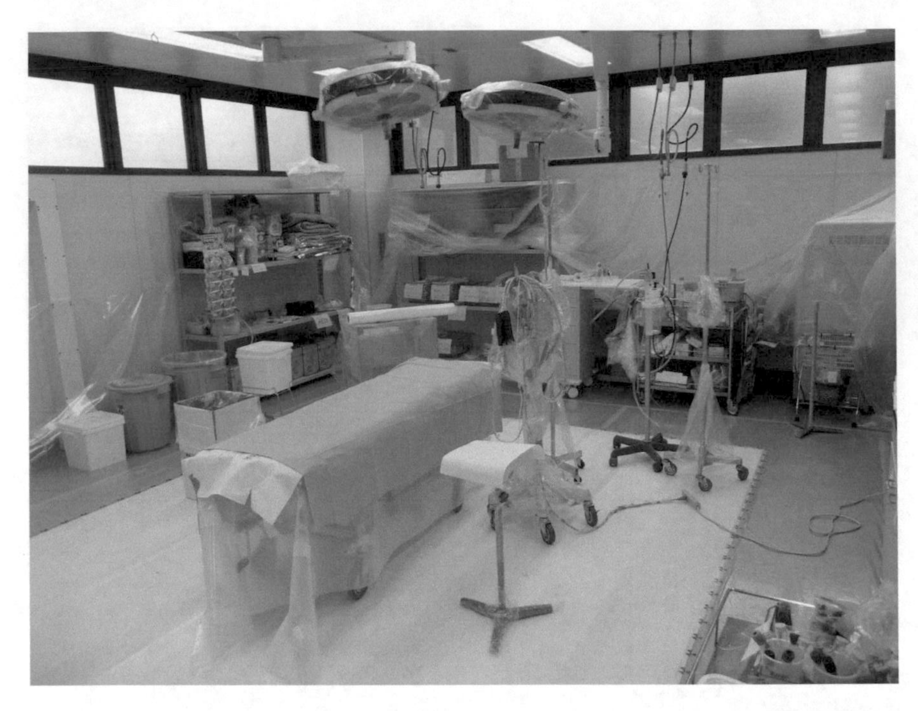

Fig. 17.4 A trauma bay at the Fukushima Medical University Hospital where patients were examined and treated in the aftermath of the Fukushima nuclear reactor accident. Note the seamless floor covering, the absorbent paper, and plastic coverings to limit contamination. Author's photo

seamless floor covering that extends at least 15 cm up the walls and should be located close to an entrance or should have an entrance directly into the room to minimize tracking contamination through the hospital. The photo below, from the Fukushima Medical University (Fig. 17.4) shows a well-designed treatment room that was used to treat contaminated patients in the aftermath of the 2011 reactor accident and Fig. 17.5 shows one example as to how a trauma bay for contaminated patients can be designed. In the photo, note the plastic coverings that help to keep furniture and equipment from becoming contaminated (the coverings can be removed and replaced quickly between patients) and the plastic-backed absorbent on the floor beneath the examination table. And while this trauma bay was not designed from the idealized radiological trauma bays described by the NCRP [24] or the American College of Radiology [2], all of these organizations came to very similar designs.

If it is not possible to have a trauma bay with a dedicated entry or that is adjacent to an entry then it will be necessary to transport contaminated patients through the Emergency Room and along corridors to reach the contaminated patients' trauma bay. In a case such as this the hospital should consider measuring and pre-cutting floor coverings from textured plastic that can quickly be laid down and taped in place; when no longer needed the plastic sheeting can be rolled up and the floor beneath surveyed for contamination. This approach was used on the nuclear submarine on which the author was stationed; the plastic sheeting could be laid out and taped to the walls in a matter of minutes.

Note, however, that arrangements such as this must be practiced periodically so that staff know where the supplies are stored and how they are to be deployed [21–23]. In fact, the most common comment made to the author by workers at the hospitals and shelters his group visited in Japan during his 2011 trip was "We knew what to do because we practiced this every year but we never expected to have to do it for real."

Putting this together, there are several steps that a hospital can take to prepare to receive contaminated patients:

1. Construct or modify at least one examination and treatment room to be suitable for receiving contaminated patients.
2. Obtain contamination control materials for additional treatment rooms.
3. Develop institutional procedures for examining and treating contaminated patients, including triage for radiological and non-radiological concerns.
4. Train staff, including periodic refresher training and exercises to develop and maintain skills and familiarity with procedures and equipment.
5. If a radiological incident occurs, set up examination and treatment rooms to receive contaminated patients.
6. Ensure staff are properly garbed in Personal Protective Equipment while waiting for the first patients to arrive.

17.3 Medical Care in the Emergency Department

Practitioners of emergency medicine have ample experience in their specialty and it is not the place of this book to go into the details of their work. However, most who practice emergency medicine know very little about radiation, its health effects, its treatment, or even how to recognize radiation injuries and, without such knowledge, it is difficult to assess the entirety of a patient's medical needs. Thus, here we will take a look at the clinical impact of radiation exposure during the first few to several hours after exposure.

17.3.1 Contaminated Wounds and Injuries and Embedded Source Fragments

Contamination, whether on the skin or in wounds, can be identified and located by performing a survey using radiation detectors. Embedded fragments can also be identified using radiation surveys, in addition to leaving entry wounds at the point where they entered the body. A high-activity source fragment within the body, though, can generate a radiation field capable of swamping contamination instruments; thus, a single source pellet of, say, a few tens of GBq can cause the entire patient to appear to be contaminated and can even cause instruments to overload due to dead time.

Contaminated and radioactive materials should be removed from the body when found. However, it might not be immediately evident which fragments are radioactive and which are not; accordingly, each object removed from the body should be removed using forceps, hemostats, or similar tools and should be surveyed and placed in a shielded container if contaminated or radioactive. Similarly, wipes and fluids used for decontamination or that are used to flush injuries or burns should be collected for later disposal as radioactive waste.

Internal radioactivity can remain in the body for weeks, months…even decades, depending on the radionuclide(s) and their chemical form. Treatment goals, then, must include cleaning wounds to prevent contamination from entering the body through breaches in the skin or from being absorbed through the skin into the body. For the most part, radioactive contamination can be cleaned from wounds and the skin the same as any other contaminants; flushing with saline solution, using cleaning wipes, debridement, and so forth. When cleaning non-injured parts of the body, care must be taken to not let contaminated liquids flow over injuries, contaminating them [24]. Wounds and burns can also be flushed with DTPA or other chelating agents to remove metals such as cesium, cobalt, americium, and so forth [13].

If contamination becomes fixed to the skin, medical personnel (possibly in conjunction with radiation safety professionals) must determine if the contamination levels are sufficiently high as to require further decontamination and, if so, what sort of decontamination is called for. Minor levels of contamination, for example, might not cause enough radiation exposure and might not pose a sufficiently high risk of absorption to require any further efforts. In the aftermath of the Fukushima accident, for example, persons with up to about 100,000 counts per minute were released from survey stations without any further decontamination requirements, as were patients admitted to local hospitals [1].

If contamination levels and/or the potential dose to the skin (or to internal organs if the contamination is absorbed into the blood) are high enough to require further decontamination, options include:

- Continued use of cleaning solutions and/or chelating agents until contamination levels are acceptable
- Skin abrasion
- Chemical removal using a compound appropriate for the contaminating element (see, e.g. [13])
- Excision and/or debridement
- Administration of decorporation agents to treat materials absorbed into the body.

Any patients with embedded fragments or contaminated wounds or burns should be assessed for internal contamination as described in Chap. 18. If an intake is found to have occurred, decorporation should be considered if the amount of uptake is greater than one clinical decision guideline—CDG—NCRP [24].

17.3.2 *Assessing the Patient's Acute Radiation Exposure and Intake*

In particular, medical caregivers must be able to determine if a patient's radiation exposure—internal or external—was high enough to cause clinical issues. There are a variety of symptoms that can indicate exposure to radiation exposures of 1 Gy or more as an acute exposure: vomiting, diarrhea, headache, and fever. The amount of time required for the onset of these symptoms can help medical caregivers to assess the acute radiation exposure; these can be compared to the dose determined from lymphocyte depletion for those who are admitted to the hospital. These are summarized in Table 17.1, adapted from Guskova et al. [11].

Evaluating these (and other) factors is most easily and reliably done using the Radiation Emergency Medical Management website's biodosimetry tools (https://www.remm.nlm.gov/ars_wbd.htm), which can be downloaded to a computer or as a smart phone app.

Assessing internal exposure is more challenging, but it is possible. This is discussed in Chap. 18 and in greater detail in several papers and reports (e.g. [4, 19, 25]). For the purposes of this chapter, let it suffice to note that patients arriving in the Emergency Department might have been scanned in the field or in a CRC, these readings should be confirmed by surveying the lungs and face and evaluating the readings as discussed in the references noted above.

Table 17.1 Acute radiation sickness doses and symptoms

Symptom	ARS degree and approximate whole-body exposure (Gy)			
	Mild (1–2 Gy)	Moderate (2–4 Gy)	Severe (4–6 Gy)	Very severe (6–8 Gy)
Vomiting	2 h after exposure	1–2 h after exposure	<1 h after exposure	<30 min after exposure
Diarrhea	None	None	Mild, 3–8 h after exposure	Heavy, 1–3 h after exposure
Headache	Slight	Mild	Moderate (4–24 h after exposure)	Severe (3–4 h after exposure)
Consciousness	Unaffected	Unaffected	Unaffected	May be altered
Fever	None	Slight (1–3 h after exposure)	Moderate (1–2 h after exposure)	High (<1 h after exposure)
Medical care	Outpatient observation	Observation in general hospital	Treatment in specialized hospital	Treatment in specialized hospital

17.3.3 Radiation Safety in the Emergency Department

The primary radiation safety concern in the Emergency Department are contamination control, followed by limiting radiation exposure from embedded source fragments.

Limiting radiation exposure from embedded source fragments requires, first, locating the fragments and obtaining an accurate dose rate. Fragments should not be held with the hands or fingers—only with forceps, hemostats, or the like—and they should be placed into a shield (colloquially called a "pig"), a shielded safe, or a related location. If removing an embedded source fragment is expected to require an extended period of time, those involved in the procedure should take care to utilize the principles of Time, Distance, and Shielding (discussed below) to reduce their own radiation exposure and that of others in the trauma bay or operating room in which the procedure is being performed (Fig. 17.5).

Contamination control is somewhat more challenging because contamination can spread quite easily from person to person, from person to objects or surfaces, and from objects or surfaces to workers. The following precautions will help to control the spread of contamination while a patient is in the Emergency Department [2, 24].

- Medical staff should assume that all patients arriving from and downwind of the site of the attack are contaminated and should take appropriate precautions.
- Uncontaminated patients should not be put into the same rooms as those who are contaminated.
- Use disposal equipment (e.g. thermometer covers, blood pressure cuffs, etc.) when possible.
- Remove patient's outer clothing if possible and place in a plastic bag.
  - Patients' clothing should be saved for possible forensic purposes; when no longer needed for that purpose it may be disposed of as radioactive waste.
- This includes taking standard precautions when working with patients as well as changing PPE and surveying one's hands after finishing and before moving on to the next patient.
- If the patient is in an examination room or bay there should be a waste container at the entry so that staff can place their used (and possibly contaminated) PPE in the container upon exiting.
- If resources permit there should be contamination survey meters (preferably a pancake-type GM) in each area so that those exiting a room or bay can survey their hands and feet for contamination prior to exiting.
- If resources permit, there should be a step-off pad or entry/exit area set up to provide a location for donning and doffing PPE, conducting contamination surveys, and performing decontamination (if necessary).
- Establish boundaries between contaminated and uncontaminated areas.

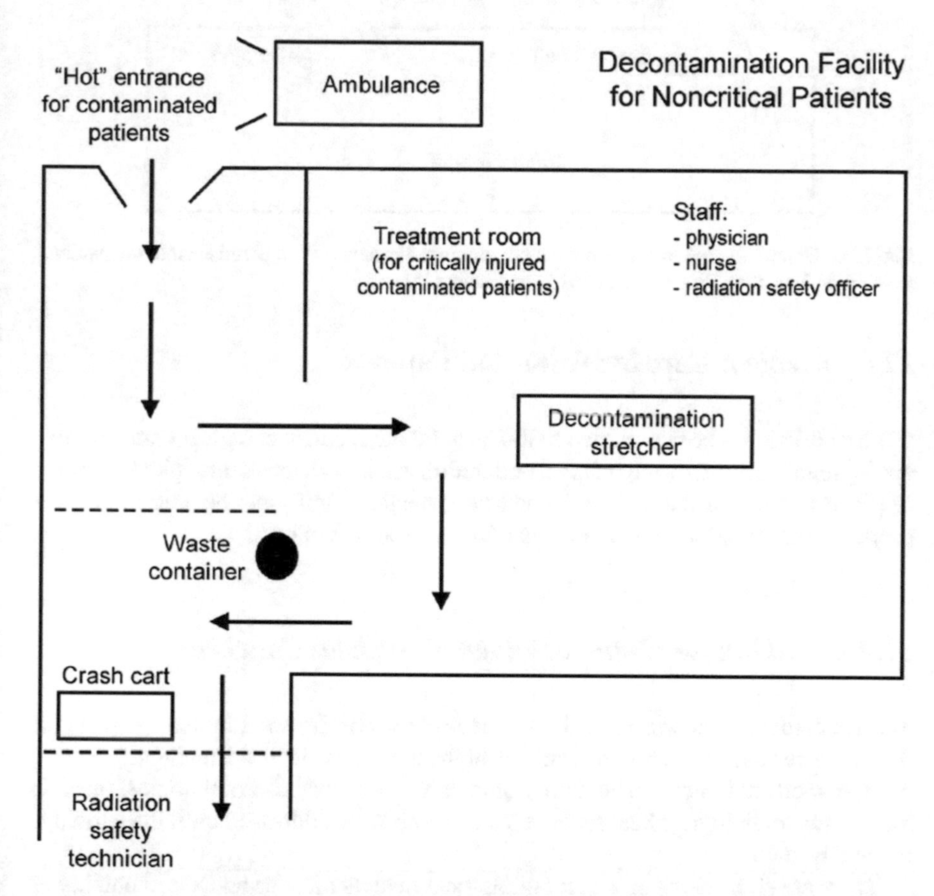

Fig. 17.5 Idealized layout for a trauma and treatment bay designed for contaminated patients. With permission of the National Council on Radiation Protection and Measurements [24]

- Survey the "cold" side of these boundaries periodically to confirm that contamination is not spreading.
- Adjust boundaries and/or decontaminate as necessary if contamination is found to have spread from the "hot" to the "cold" side of the boundaries.

- Ensure that all patients are decontaminated before they are released from any "hot" area, whether for discharge or to be admitted (Fig. 17.6).

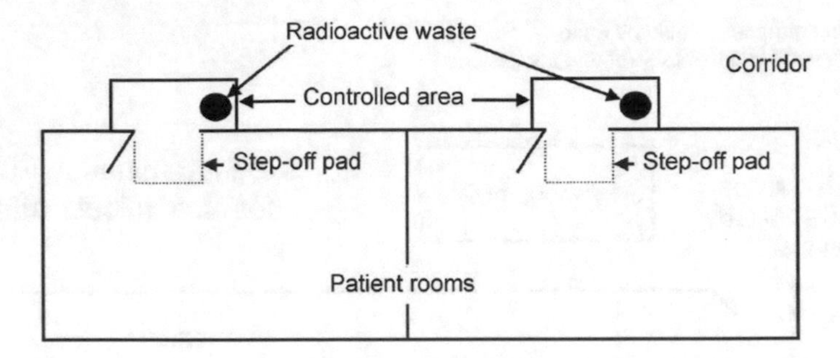

Fig. 17.6 Entry/exit area for the rooms of contaminated patients. With permission of the National Council on Radiation Protection and Measurements [24]

17.4 Medical Care for Admitted Patients

Once a patient has been admitted to the hospital the medical caregivers must be able appropriately prioritize and balance both radiological and non-radiological concerns to properly care for the patient. In addition, medical staff must be able to practice proper radiation safety. Each of these will be discussed in turn.

17.4.1 Addressing Non-radiological Medical Concerns

Serious medical concerns should take the highest priority for admitted patients for the same reason they should take the highest priority in the Emergency Room; because medical issues can be more immediately fatal than can radiological ones. In most cases, radiological health concerns can wait to be addressed until the patient is medically stable.

There are some cases in which radiological matters must be addressed quickly as part of stabilizing the patient; embedded source fragments, for example, can cause extensive tissue damage if not removed promptly, while some radionuclides are best removed from the body when decorporation therapy begins quickly. These will be discussed in the following section. Aside from situations such as these, medical caregivers should use their medical judgement to prioritize care for their patient. In the case of a patient who has inhaled potentially dangerous amounts of alpha radioactivity (ten CDG or more), consideration should be given to preparing the patient for pulmonary lavage to remove the radioactivity from the lungs before it can cause too much damage or be absorbed into the blood.

A number of samples should be obtained as soon as practicable after admission [18, 34]:

- **CBC and differential** immediately followed by absolute lymphocyte counts every 6 h for 48 h if whole-body irradiation is considered possible
- **Routine urinalysis**

- **Swab body orifices** if external contamination is suspected
- **Swab wounds or measure count rate from dressings** if external contamination is suspected
- **24-h urine collection** for four days if internal contamination is suspected
- **24-h fecal collection** for four days if internal contamination is suspected.

High levels of radiation exposure can cause additional complications such as skin burns, blistering, hair loss, and so forth; in general these are unlikely to appear within a few to several days after exposure and they can usually be treated by treating the symptoms. This is discussed in greater detail in the following section. Medical caregivers will also need to take appropriate radiation safety precautions to protect themselves and others under their care; these are also discussed below.

17.4.2 Addressing Radiological Health Concerns

Radiological health effects all depend on the amount of radiation exposure a person has received; deterministic health effects depend on the acute radiation exposure and stochastic health effects are a function of radiation exposure over a lifetime. Therefore, an estimate of a patient's radiation exposure from both internal and external radioactivity is important to know or to estimate. If it can be determined that a patient has less than 1 CDG of internal contamination and external exposure is estimated to have been less than 0.25 Sv then it is likely that radiation exposure will have little or no impact on a patient's clinical presentation. At higher levels of exposure, the radiation exposure becomes increasingly important; at exposures (internal and external combined) of about 3 Sv, not only does the radiation begin to cause short-term health effects (e.g. skin burns) but it can also be fatal to some of those exposed.

The progression of symptoms in admitted patients can help physicians to determine the likely radiation exposure and the patient's future prognosis. Table 17.2 lists a number of symptoms and their time to onset among acutely exposed patients (from [11, 15]).

Some of the deterministic effects can be treated symptomatically. The author had the opportunity to meet the physician who treated the skin burns of two Fukushima workers whose boots filled with radioactive water while they were inspecting the basement of a building following the accident. The physician mentioned that the two workers had second-degree burns over their lower legs, and that he treated these the same as any other second-degree burns; the fact that they were caused by radiation did not affect his treatment plans. It must be noted that erythema, epilation, blistering, ulceration, and necrosis are all injuries that require doses of 3 Gy or more to the skin and that these do not manifest themselves for at least 2 weeks post-exposure. Thus, any burns that are seen in the Emergency Department on the day of a radiological attack are likely to be due to thermal, chemical, or even electrical damage to the skin and not due to radiation.

Table 17.2 Degrees of acute radiation sickness and approximate whole-body dose

Symptom	1–2 Gy	2–4 Gy	4–6 Gy	6–8 Gy	>8 Gy
Onset of symptoms (days)	>30	18–28	8–18	<7	<3
Lymphocytes (G/L)	0.8–1.5	0.5–0.8	0.3–0.5	0.1–0.3	0.0–0.1
Platelets (G/L)	60–100	30–60	25–35	15–25	<20
Clinical manifestations	Fatigue, weakness	Fever, infections, bleeding, weakness, epilation	High fever, infections, bleeding, epilation	High fever, diarrhea, vomiting, dizziness, disorientation, hypotension	High fever, diarrhea, unconsciousness
Lethality	0%	0–50% (6–8 weeks)	29–70% (4–8 weeks)	50–100% (1–2 weeks)	100% (1–2 weeks)
Medical response	Prophylactic	Prophylactic (days 14–20), isolation (days 10–20)	Prophylactic (days 7–10), isolation from first day	Special treatment, isolation from first day	Symptomatic only

For those interested in reading specific case studies, Gusev et al. [10], Ricks et al. [30], and a number of IAEA reports include numerous case studies that reward a careful reading.

17.4.3 Good Radiological Work Practices

Radiation workers receive extensive training on good work practices, training which includes practical factors and hands-on experience in proper PPE and how to don and doff the equipment, various aspects of contamination control, performing radiological surveys, various decontamination techniques, and much more. Medical personnel, however—even those who work in Radiology, Nuclear Medicine, and Radiation Oncology—receive far less radiation safety training; the majority receive no such training whatsoever. Because of this, it is not uncommon for medical personnel to feel uncomfortable around, or to be frightened of radiation and radioactivity; this can extend to patients who are contaminated with radioactivity or who are brought in from contaminated areas. This is exacerbated by a relative paucity of knowledge about the health effects of radiation on the part of most medical caregivers [31]. In fact, concerns about the health effects of radiation are one of the reasons that multiple surveys have indicated that as many as half of medical responders would decline to go to work following a radiological or nuclear attack, mostly due to fears of the health effects of radiation [3].

Patients in the Emergency Department are likely to remain for only a relatively short period of time and, while there will be a need to exercise caution with patients who are contaminated or who have embedded radioactive fragments, there are different concerns in the Emergency Department compared to the concerns regarding working on a patient who is admitted for what might be an extended period of time.

Much of good radiological work practices revolves around controlling radiation exposure and controlling the spread of contamination. Each of these will be addressed in turn.

Health Physicist Strom [32] published *The Ten Principles and Ten Commandments of Radiation Protection.* While not all of Strom's principles apply in a medical setting, there are a number that do:

- **Time**—minimize the amount of time spent working in the highest-dose radiation fields and do not enter such areas unless necessary. When working in any radiation area, work quickly (although not hastily) and leave as soon as the necessary work is finished. Have conversations and make radio or telephone calls in areas that have low dose rates if possible. When examining or treating a patient, minimize the amount of time spent in close proximity to the patient.
- **Distance**—because of the inverse square law radiation dose rates drop off quickly with distance; doubling the distance to a radiation source will reduce the dose by a factor of four. This is true at close distances as well as on larger scales. For example, when attending to a heavily contaminated patient, a patient with an intake of radioactivity, or a patient with embedded source fragments, standing at arms' length instead of immediately next to the patient can reduce exposure. If the patient is known to have radioactive material in a specific location (e.g. I-131 in the thyroid, source fragments embedded in the leg, etc.) try to stand away from that part of the body—stand near the head of a patient with fragments embedded in the calf, stand at the foot of the bed for patients with an I-131 intake.
- **Shielding**—interposing radiation shielding between the caregiver and the radiation source will also help to reduce exposure. Many hospitals with radiation oncology programs, for example, will have rolling or mobile shields that can be used for patients with embedded source fragments.
- **Source reduction**—flushing wounds with saline solution or a chelating agent (e.g. DTPA) not only reduces intake but will also reduce the amount of radioactivity to which personnel are exposed. This would include removing embedded source fragments (if any), which should then be placed into shielded containers (*shielding*) or moved to a radioactive material storage area (*distance*) until they can be sent for disposal.
- **Surveys**—to help identify areas with elevated levels of radiation, radioactive sources, and areas with contamination. This should include hand and foot surveys for everyone leaving the room of a contaminated patient as well as using step-off pads and performing contamination surveys of the floor outside their rooms.
- **Decorporation** (if appropriate)—treating those with an intake of radioactivity will reduce their exposure.

- **Personal protective equipment**—use of appropriate PPE will help to keep contamination from entering the body as well as reducing dose to the skin from contamination. Wearing surgical gloves and gowns or other protective clothing when entering the rooms of contaminated patients is one example of appropriate protective gear. PPE should also include dosimetry where possible, to monitor radiation exposure so that workers know how much exposure they have received.
  - It is important to note that the hospital workers caring for Alexander Litvenenko did not know that he had been poisoned with radioactivity and was shedding radioactivity during the three weeks of his hospitalization. In spite of this, there were no workers who received a significant intake of Po-210; the reason for this is that the standard precautions taken with all patients suffering from unknown illness (Universal Precautions, also known as Standard Precautions) were sufficient to limit exposure to contamination.

There is more, though, that can be done beyond these. Contamination control, for example, begins before the first patient arrives at the Emergency Room doors—it begins with controlling contamination on the patients (if their medical condition permits) before they are brought to the Emergency Room and with setting up the Emergency Room to receive contaminated patients as discussed in the first part of this chapter.

In addition to the above, medical personnel should also consider taking additional actions to help control the spread of contamination. These include:

- Removing protective clothing upon leaving one room and donning new and uncontaminated protective gear prior to entering the next.
- Establishing contamination control areas at the entry to each patient's room.
- Using proper procedures for entering and exiting the rooms of patients known to be contaminated, as described in NCRP Report 161 [24].

17.5 Decorporation of Internal Radionuclides

Internal radioactivity can produce radiation dose for years or decades after intake. Inhaled highly insoluble plutonium [20], for example, can remain in the lungs for years, once it enters the bone it will remain for decades, as will americium, and other actinide elements. But if a medication called Diethylenetriamine pentaacetate (DTPA) is administered to the patient the amount of plutonium, americium, and other actinide elements remaining in the body is reduced considerably with a corresponding reduction in radiation dose over the years. Depending on the radionuclide in question and the agent used this sort of therapy can reduce radiation exposure considerably—the administration of potassium ferrocyanide (popularly known as Prussian blue) to those exposed to Cs-137 during the accident in Goiânia Brazil increased the Cs-137 excretion by a factor of three with a concomitant reduction in radiation exposure to those receiving the medication [9].

The process of removing internal radioactivity is called decorporation and there are a number of agents that can be used to decorporate many radionuclides to which a person might be exposed. It must be noted that a decorporation agent will act against all nuclides of a given element, not only against those that are radioactive. Thus, if a person had an uptake of, say, radioactive cobalt and a decorporation agent were administered, it would act to scavenge both radioactive cobalt to which a person had been exposed as well as the stable cobalt present in the heart, liver, pancreas, that is used in the synthesis of Vitamin B^{12} and any of a number of proteins and enzymes used throughout the body. Because all cobalt isotopes are chemically identical, it is not possible to decorporate only the radioactive isotopes.

Some decorporation agents are so simple to administer that this can be done by the patients themselves. Potassium iodide (KI), for example, can be provided in advance of any exposure and the administration instructions can easily be printed on the side of the bottle of tablets [26]. In fact, many states in the US pass out KI to all citizens living within a given distance of a nuclear power plant so that, in the event of a meltdown, public health officials can order everybody to take the KI they have at their homes. Potassium ferrocyanide (Prussian blue), a compound that would be taken after exposure to radioactive cesium, is also relatively easy to administer and capsules can be taken by patients. These are also safe enough that, even if the patients accidentally misadminister the medication, they are likely to be unharmed.

Other decorporation agents are either more difficult to administer or must be administered more carefully to avoid toxicity concerns. Diethylenetriamine pentaacetate (DTPA), for example, is typically administered intravenously and can cause chills, diarrhea, fever, and other effects; it can also scavenge zinc from the body, requiring zinc supplements [29]. As if that were not enough, there are different forms of DTPA—calcium and zinc—each with its own requirements and precautions for administration. The Food and Drug Administration also recommends caution when administering Ca-DTPA to patients with hemochromatosis (a rare genetic condition that affects the manner in which our bodies process iron). As a more complex medication, DTPA requires more care and more skill from those prescribing and administering it; unlike KI, we cannot hand out DTPA in advance or in anticipation of an accident or attack.

Other documents (e.g. [10, 24, 25, 29]) go into detail with regards to dosing, counterindications, precautions, and the like and those will not be repeated here. Table 17.3 was developed using information from the references above and provides a summary of some relevant information for a number of medical countermeasures, but it is not intended to be comprehensive. Decorporation is a part of the overall medical management of patients that have ingested or inhaled radioactive materials, and it must be coordinated with the medical personnel attending to the patient's other injuries [5].

Table 17.3 Nuclides, decorporation agents, and notes on proper administration

Nuclide	Target organs	Agent	Administration route	Considerations and comments
Americium (Am-241)	Lungs, liver, bone, bone marrow	DTPA (chelating agent)	Intravenous	Begin with Ca-DTPA and change to Zn-DTPA if necessary. Can use DTPA solution to flush contaminated wounds
Cesium (Cs-137)	Whole body	Prussian blue (chelating agent)	Capsule—by mouth—taken for minimum of 30 days	Obtain bioassay prior to starting treatment and periodically during treatment
Cobalt (Co-60)	Liver	Succimer (chelating agent)	Capsule by mouth every 8 h for 5 days, then every 12 h for 2 weeks	
		DTPA (chelating agent)	Intravenous—continue until body burden is acceptable	Can use DTPA solution to flush contaminated wounds
		EDTA (chelating agent)	Intravenous (single dose) or injection (divided dose)	If injected, give doses 8–12 h apart
Iodine (I-131, I-133)	Thyroid	Potassium iodide (blocks uptake)	Pill—by mouth—taken until environmental iodine has decayed or patient moves to uncontaminated location	Not recommended for adults older than 40 or for those with allergies to KI; lower dose for children
Iridium (Ir-192)	Spleen	DTPA (chelating agent)	Intravenous—continue until body burden is acceptable	Can use DTPA solution to flush contaminated wounds
		EDTA (chelating agent)	Intravenous (single dose) or injection (divided dose)	If injected, give doses 8–12 h apart
Polonium (Po-210)	Spleen, kidneys, lymph nodes, bone marrow, liver, lung, mucosa	Gastric lavage		Must perform procedure while material is still in the stomach
		Dimercaprol (British Anti-Lewisite)	Injection (IM) for 10 days	

(continued)

Table 17.3 (continued)

Nuclide	Target organs	Agent	Administration route	Considerations and comments
		Succimer (chelating agent)	Capsule by mouth every 8 h for 5 days, then every 12 h for 2 weeks	Typical course of treatment is about 19 days
		Penicillamine (chelating agent)	Capsule by mouth	Perform bioassay to monitor efficacy, continue as long as clinically indicated
Strontium (Sr-90)	Bone	Calcium gluconate (phosphate binder, competes for binding sites on bone)	Intravenous for 6 days or longer as needed	May also include stable Sr compounds
		Barium sulfate (blocks intestinal absorption)	Mix with water and drink—one dose to block absorption by intestine	Must be administered before absorption can occur

17.6 Community Reception Centers

In the aftermath of the radiological accident in Goiânia Brazil there were over 110,000 people who came to the city's soccer stadium to be surveyed for contamination [14]. Of those, about 250 required decontamination and a further 125 required some form of medical attention. Those desiring a survey comprised about 10% of the population of Goiânia; if this is extrapolated to New York City then the number of people desiring evaluation might well be more than one million, and even smaller cities such as Chicago and Los Angeles would be trying to survey several hundred thousand people if we include tourists, commuters, and those living in nearby suburbs who work in the city.

If a person is worried about their health, the tendency is for them to see a physician; if even just 10% of a city's population descends upon its hospitals seeking to be scanned for internal and external radioactive contamination, a hospital (or multiple hospitals) can quickly become so clogged with those who are worried (albeit healthy) that they can lose the ability to care for those with broken bones, severe lacerations, heart attacks, and other medical problems. As a result, many cities are planning on standing up Community Reception Centers (CRCs) that are intended to scan at least 1000 people every hour and to identify and decontaminate those who have an elevated count rate, as well as to identify those whose contamination is internal and must be decorporated rather than simply cleaned up.

The basic idea of a CRC is to have a facility that is dedicated to—and optimized for—high throughput. Several major American cities have developed CRC plans that call for multiple CRCs, each capable of surveying as many as 1000 people every hour, working around the clock so that about one million people could be surveyed each week. But CRCs would not be stood up simply to survey people—they would also be expected to identify those with external contamination and to decontaminate them as well as identifying those with internal contamination who require decorporation therapy [7]. A diagram of an idealized CRC showing the major sections is shown in Fig. 17.7.

The CRC should have "hot" and "cold" sides—anti-contamination clothing, dosimetry, and other protective measures will be required on the "hot" side and they may not be necessary on the "cold" side, provided that nobody is permitted to enter the cold side until they have been either surveyed or decontaminated. However,

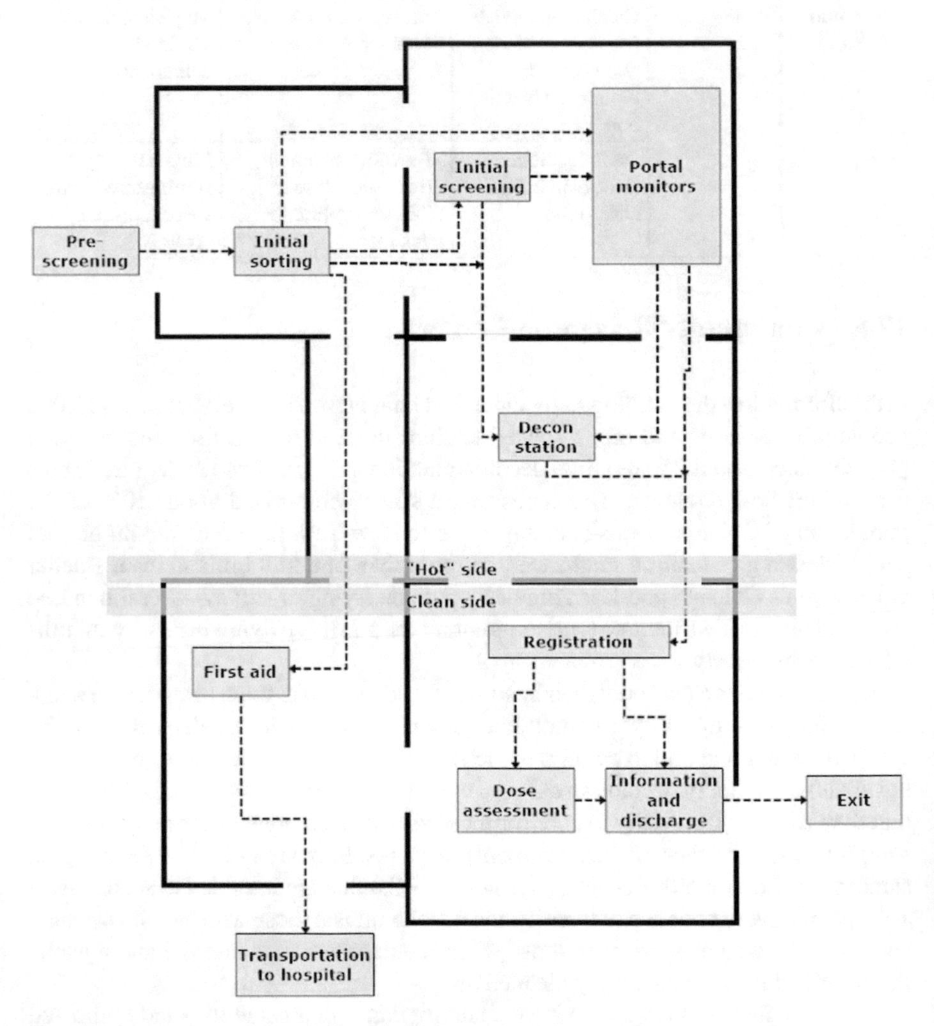

Fig. 17.7 An example of how a Community Reception Center might be organized

it will be necessary to perform contamination surveys and to pull air samples regularly on both the hot and cold sides as long as the CRC is in operation in order to confirm that the "cold" side remains contamination-free. The boundary between the hot and cold sides will be the screening station.

As envisioned, a CRC will include several stations, described below:

1. **Preliminary screening and sorting**—as members of the public are lined up waiting to enter the CRC personnel will circulate among them performing radiation surveys. Anybody who is emitting elevated levels of radiation will be taken aside to determine if they are nuclear medicine patients or if they were contaminated and require decontamination. Pulling people out of the line at this point will get them decontaminated more quickly (if they are contaminated) and will keep a single person from triggering alarms on as many as ten portal monitors simultaneously when they walk into the screening room. During this time, those waiting in line might also be given information cards to fill out and/or shown videos explaining what to expect as they pass through the CRC.
2. **First aid**—while public service announcements will encourage those who are injured to seek medical attention, it is likely that there will be many who do not consider their injuries to be sufficiently serious as to warrant medical attention, and it is possible that there will be others who fall ill or who are injured while at the CRC. The First Aid station will be responsible for caring for these people and for identifying those who require more advanced medical care at a hospital or urgent care center.
3. **Whole-body screening**—Many CRC plans call for screening to be performed using multiple walk-through portal monitors in a large room such as a gymnasium. Those being screened would be directed into one of ten lanes (more or less, depending on the specific CRC design) where they will approach and walk through the monitor. If they have internal or external contamination the monitor will alarm and they will be sent to the decontamination station. The challenge at the screening station will be to set portal monitor alarms low enough so that those with more than one CDG of internal contamination will trigger the alarms, while keeping alarm setpoints high enough to minimize false alarms.
4. **Decontamination**—Those sent to the decontamination station will remove their outer clothing and will be decontaminated using an appropriate methodology (wet, dry, moist, or a combination). Decontamination will continue until the person is decontaminated (the count rate on the person is below acceptable levels) or until it is clear that the skin is decontaminated and the remaining contamination must be internal. These groups will both be sent to the registration station, and the second group will then proceed to the dose assessment station.
5. **Radiation dose assessment**—those who meet certain criteria—typically having evidence of internal and/or external contamination—will be sent to the dose assessment station. The person(s) staffing the dose assessment station will review the information collected and might collect additional information; one of the dose assessment methods discussed in Chap. 18 will be used to determine

whether or not the amount of internal contamination warrants decorporation therapy, further tests, or no action.

6. **Registration**—Every person passing through the CRC will be required to stop at the registration station. Here their personal information, contact information, and survey information will be recorded; this information will be used for long-term tracking and follow-up, especially for those found to have internal and/or external contamination. Information recorded at the registration station will be used for long-term tracking and follow-up of the health of those registered as well as for possible scientific research and for adjudicating future compensation claims among those who develop cancer at any time after the attack.
7. **Mental health**—any terrorist attack is likely to traumatize members of the public and to disrupt the lives of those in the city that was attacked. The mental health station will help these people to begin to recognize the shock they have undergone and to help them to find the resources and the help they will need.
8. **Information and discharge**—Before going home, all persons will have the opportunity to learn about shelters and assistance programs relevant to their needs. Some shelters might require proof that a person is not contaminated in order for them to be admitted; the information and discharge station should also be able to provide them with a certificate indicating their contamination status.
9. **Investigations**—some of those passing through the CRC might have witnessed events that could be important to investigating the attack; it is also possible that those who perpetrated the attack might come to one of the CRCs for assessment and (possibly) decontamination. An investigations station will be established to interview those who might have information that could be of value to the criminal investigation. In addition, some at the CRC will be trained to recognize radiation injuries and will bring these to the attention of the investigations section if they are noticed.

There is more to the CRC than the public passing through; it will be necessary to set up the CRCs and make them operational, as well as to operate them for weeks or even months following an attack. This means that each CRC will require a management structure, a Health and Safety Officer, a Radiation Safety Officer, and two or three shifts of staff for each station as well as for operating the CRC as a whole. As one example, it will be necessary to take periodic air samples, to perform periodic contamination surveys on both the "hot" and "clean" sides of the CRC, to collect contaminated clothing and personal effects, and all of the other functions that will be required.

Operating a CRC requires a large number of people as well; the following numbers are estimates based on the type of CRC envisioned by several large cities.

- Managers: 5–10 staff
  - CRC Manager, Radiation Control Officer, Safety Officer, Operations Manager, HazMat Operations Director, etc.
- Pre-screeners: 2–5 staff
- First Aid station: 2 staff

- Whole-body screening: 40–50 staff
  - 2 for each portal monitor,
  - 20 for flow control,
  - 10 monitoring with hand-held detectors
- Decontamination: 20 staff
- Radiological Assessment: 10 staff
- Registration: 100 staff
  - Gives 10 min per person passing through CRC
- Mental health: 20 staff
- Information and discharge: 20 staff
- Investigations: 4 staff
- Miscellaneous: 20
  - Security, cleaning, radiological surveys, etc.

This is about 250 people per shift for each CRC; a six-CRC system such as the one planned by New York City will require close to 3000 people to operate for two shifts daily. More will be needed if CRC staffing plans include a third shift or to allow for people to take weekends off. Given the large number of people and skills required to operate a CRC it is reasonable to assume that multiple agencies will be involved, along with suitably trained volunteers or paid non-government assistants.

Given the large number of people filling a variety of positions it will be important to determine which agency is responsible for filling which positions or staffing specific stations. It will also be vitally important to develop appropriate procedures, policies, forms, and training to ensure that those filling these positions are able to do so competently.

In addition to the administrative preparation noted above, it is also important to develop an inventory list for all of the equipment and supplies needed to set up and operate a CRC for up to one week. In New York City, for example, the Fire Department maintains six trailers, each filled with the radiation detectors, PPE, boundary rope, floor coverings, and the other materials that will be needed to operate the CRC for at least the first few days—remembering that a facility designed to process 1000 people per hour will need sufficient supplies to attend to nearly 50,000 people in the first two days alone.

Any city that plans to stand up and operates one or more CRCs will need to do extensive planning in advance, will need to accumulate the supplies required to operate these facilities, and will need to not only identify, but also to train those who will operate them. This process can be lengthy and expensive; New York City spent over five years and nearly $1 million developing their CRC plans and purchasing the supplies for their CRC "kits."

17.7 Radioactive Decedents

In any other than the most minor attacks people are going to die, whether killed by the blast or dying later from their injuries. It is inevitable that some decedents will be contaminated or might even contain embedded radioactive fragments and it is important that they be handled both respectfully and safely. To that end, guidance on this matter has been developed [6, 24], to help physicians, medical examiners, and mortuary workers to do their work while remaining safe.

The NCRP report #161 [24] begins with a reminder that any bodies found at the scene of an attack should not be moved or decontaminated, and their clothing or personal effects should not be examined until they have been photographed and examined by the Medical Examiner or members of the Medical Examiner's staff. They also recommend labeling the bodies and body bags with the radiation symbol and, if possible, storing them in a field morgue to minimize contamination of facilities, although those that have low levels of contamination can be sent to fixed facilities for examination.

The primary radiation safety concerns are radiation and contamination from the bodies being examined. Contamination can place constraints on the manner in which work is done; specifically, high levels of contamination can call for workers to wear appropriate personal protective equipment and to need to survey themselves, as well as requiring surveys of the body and the examination room during and after each autopsy. But contamination in and of itself does not normally pose a health risk. The CDC also notes that placing the body bag inside a plastic container after removing it from the scene can help to contain any contamination that might be on the exterior of the body bag.

Radiation is not normally expected to be dangerously high from contamination alone, but embedded fragments of radioactive materials might be. For example, consider a Co-60 source set on top of high explosives. Cobalt-60 sources are usually comprised of a solid piece of metal alloy; when the metal is hit with a shock wave, pressure, and high temperatures it can spall or break into multiple pieces and any of these pieces can fly off and become embedded in a bystander to the explosion. If the source fragment contains a relatively large amount of radioactivity the ensuing radiation levels can pose a risk to the medical examiner while lower-activity fragments can cause high local radiation dose to the fingers and hands of personnel working in close proximity or handling the sources. For this reason, the medical examiner (or a radiation safety professional) should perform a quick survey of any remains with elevated dose rates to locate any embedded fragments so they can be removed, shielded, and stored for future disposal.

If possible, all remains should be surveyed in the field before moving them or at the field morgue to detect the presence of any high-activity radioactive materials (e.g. sources or source fragments) that might pose a risk to medical examiners and/or mortuary personnel. Any such sources or source fragments that are found should be removed and placed into a shielded container if possible; if removing them is not possible then the remains should be labeled with the radiation symbol and, if possible, shielded to reduce radiation dose rates to less than 1 mGy h^{-1}. This survey will also make it possible to identify the presence of high levels of contamination as well as radiation dose rates that, while not hazardous, might call for stay time restrictions or other radiation safety precautions among those examining or handling the remains.

When the Medical Examiner's work has been completed the body can be released to the family for final arrangements. Some religious faiths require preparation of the body for burial and/or a family member to accompany the body; unless the body contains high-activity embedded fragments it should not pose a risk to family members or to those conducting burial preparations.

Both CDC and NCRP advise against cremation to prevent contamination of the furnace and release of radioactivity into the environment. They also note that any embedded fragments will remain in the ash and might expose crematory personnel to elevated levels of radiation. If the decision is made to scatter the decedent's ashes, CDC recommends waiting for ten half-lives if possible, although this will not be possible for longer-lived radionuclides (Co-60, for example, has a half-life of 5.27 years and a ten-half-life delay amounts to more than 50 years). Burial is preferred, ideally in a sealed metal casket placed into a plastic-lined concrete vault to delay the release of radioactivity into the environment. The CDC also notes that it would be prudent to place a discreet warning label on the casket noting the date, the presence of radiation and/or contamination, and the dose rate emanating from the body.

References

1. Akashi M (2012) Fukushima Daiichi nuclear accident and radiation exposure. JMAJ 55(5):393–399
2. American College of Radiology (2006) Disaster preparedness for radiology professionals: response to radiological terrorism, government version 3.0. American College of Radiology
3. Balicer R, Catlett C, Barnett D, Thompson C (2011) Characterizing hospital workers' willingness to respond to a radiological event. PLoS ONE 6(10):e25327. https://doi.org/10.1371/journal.pone.0025327.AccessedNovember22,2020
4. Bolch W, Hurtado J et al (2012) Guidance on the use of handheld survey meters for radiological triage: time-dependent detector count rates corresponding to 50, 250, and 500 mSv effective dose for adult males and adult females. Health Phys 102(3):305–325
5. Breitenstein B (2003) The medical management of unintentional radionuclide intakes. Radiat Prot Dosimetry 105(1–4):495–497
6. Centers for Disease Control and Prevention (2007) Guidelines for handling decedents contaminated with radioactive materials. CDC, Atlanta
7. Centers for Disease Control and Prevention (2014) Population monitoring in radiation emergencies, 2nd edn. CDC, Atlanta
8. Coleman C, Adams S, Adrianopoli C et al (2012) Medical planning and response for a nuclear detonation: a practical guide. Biosecur Bioterror 10(4):346–371
9. Goans R (2001). Update on the treatment of internal contamination. In: Ricks RC, Berger ME, O'Hara FM (eds) The medical basis for radiation-accident preparedness: the clinical care of victims. Proceedings of the fourth international REAC/TS conference on the medical basis for radiation accident preparedness, Orlando, pp 201–216

10. Gusev I, Guskova A, Mettler F (eds) (2001) Medical management of radiation accidents. CRC Press, Boca Raton
11. Guskova A, Baranov A, Gusev I (2001) Acute radiation sickness: underlying principles and assessment. In: Gusev I, Guskova A, Mettler F (eds) Medical management of radiation accidents. CRC Press, Boca Raton, pp 33–52
12. Hick J, Weinstock D, Coleman C, Hanfling D et al (2011) Health care system planning for and response to a nuclear detonation. Disaster Med Public Health Prep 5:S73–S88
13. Ilyin L (2001) Skin wounds and burns contaminated by radioactive substances (metabolism, decontamination tactics, and techniques of medical care). In: Gusev IA, Guskova A, Mettler F (eds) Medical management of radiation accidents. CRC Press, Boca Raton, pp 363–419
14. International Atomic Energy Agency (1988) The radiological accident in Goiânia. IAEA, Vienna
15. International Atomic Energy Agency (1998) Diagnosis and treatment of radiation injuries. Safety reports series no. 2. IAEA, Vienna
16. International Atomic Energy Agency (2003) Medical management of persons internally contaminated with radionuclides in a nuclear or radiological emergency: a manual for medical personnel. IAEA, Vienna
17. International Atomic Energy Agency (2018) Method for developing arrangements for response to a nuclear or radiological emergency. IAEA, Vienna
18. Karam P (2019) Radiological incidents and emergencies. In: Veenema T (ed) Disaster nursing and emergency preparedness for chemical, biological, and radiological terrorism and other hazards, 4th edn. Springer Publishing Company, pp 569–590
19. Korir G, Karam P (2018) A novel method for quick assessment of internal and external radiation exposure in the aftermath of a large radiological incident. Health Phys 115(2):235–261
20. Ménétrier F, Grappin L, Raynaud P et al (2005) Treatment of accidental intakes of plutonium and americium: guidance notes. Appl Radiat Isot 62:829–846
21. Mettler F (2001a) Hospital preparation for radiation accidents. In: Gusev I, Guskova A, Mettler F (eds) Medical management of radiation accidents. CRC Press, Boca Raton, pp 425–436
22. Mettler F (2001b) Emergency room management of radiation accidents. In: Gusev I, Guskova A, Mettler F (eds) Medical management of radiation accidents. CRC Press, Boca Raton, pp 437–448
23. Mettler F (2001c) Application of radiation protection principles to accident management. In: Gusev IA, Guskova AK, Mettler FA (eds) Medical management of radiation accidents. CRC Press, Boca Raton, pp 449–452
24. National Council on Radiation Protection and Measurements (2008) Report No. 161, Management of persons contaminated with radioactivity, vol 1. NCRP, Bethesda
25. National Council on Radiation Protection and Measurements (2010) Report No. 166: Population monitoring and radionuclide decorporation following a radiological or nuclear incident. NCRP, Bethesda
26. National Research Council (2004) Distribution and administration of potassium iodide in the event of a nuclear incident. National Academies Press, Washington
27. Poeten R, Glines W, McBaugh D (2009) Planning for the worst in Washington State initial response planning for improvised nuclear device explosions. Health Phys 96(1):19–26
28. Radiation Emergency Assistance Center/Training Site (2020) Radiation patient treatment version 3.1. Oak Ridge Associated Universities, Oak Ridge
29. Radiation Emergency Medical Management (2020) Fact sheet: Ca-DTPA/Zn-DTPA (diethylenetriamine pentaacetate). REMM 2020. https://www.remm.nlm.gov/dtpa.htm. Accessed 22 Nov 2020
30. Ricks R, Berger M, O'Hara F (2001) The medical basis for radiation-accident preparedness: the clinical care of victims. In: Proceedings of the fourth international REAC/TS conference on the medical basis for radiation accident preparedness, Orlando
31. Shralkar S, Rennie A, Snow M, Galland R, Lewis M, Gower-Thomas L (2003) Doctors' knowledge of radiation exposure: questionnaire study. BMJ 327:371–372

32. Strom D (1996) Ten principles and ten commandments of radiation protection. Health Phys 70(3):388–393
33. Thompson J, Rehn M, Lossius H, Lockey D (2014) Risks to emergency medical responders at terrorist incidents: a narrative review of the medical literature. Crit Care 18:521–530
34. Veenema T, Karam P (2003) Radiation: clinical responses to radiologic incidents and emergencies. Am J Nurs 103(5):32–40

Chapter 18
Radiological Assessment and Public Health Response

Radiological assessment is the process of determining the amount of radiation exposure a person has received or, in the case of internal radioactivity, the amount of radiation exposure they are likely to receive in the future from the radioactivity in their bodies. The simplest and most accurate form of radiological assessment occurs when the person being assessed was wearing a dosimeter and it can be confirmed that there was no intake of radioactivity. Unfortunately, this does not happen often.

In the initial hours following any sort of radiological event it will be important to be able to determine who is likely to survive their radiation exposure, who is likely to survive with suitable care, and who cannot be saved. With limited medical resources, this process of triage will help to focus attention and resources on those most likely to benefit from that attention; to make the transition from patient-centered medicine (trying to save every single patient arriving at the hospital) to community centered medicine (spending the most time on those patients who will benefit most from the attention to save the greatest number of lives). Here, too, radiological assessment plays a huge role; unless we have some way of understanding how much radiation a patient has received (or is likely to receive) we cannot properly address the needs of the patient or the needs of the public as a whole.

Over the longer term, radiological assessment continues to play a role, especially with regards to adjudicating future compensation claims, just as assessments of exposure to airborne contaminants have been used to help determine whose lung disease might have been due to their presence in Lower Manhattan on September 11, 2001 [13]. Similarly, in the longer-term aftermath of a radiological or nuclear attack it will be helpful to have a rough guess as to the numbers of cancers and other radiation-related injuries that might arise over the years (Fig. 18.1).

One concept that can be useful is that of the Clinical Decision Guideline (CDG) first developed by the National Council on Radiation Protection and Measurements [8, 9]. As defined by NCRP, an intake of 1 CDG will expose the person affected to a committed (e.g. over the next 50 years) whole-body radiation exposure of 250 mSv;

P. A. Karam, *Radiological and Nuclear Terrorism*,
Advanced Sciences and Technologies for Security Applications,
https://doi.org/10.1007/978-3-030-69162-2_18

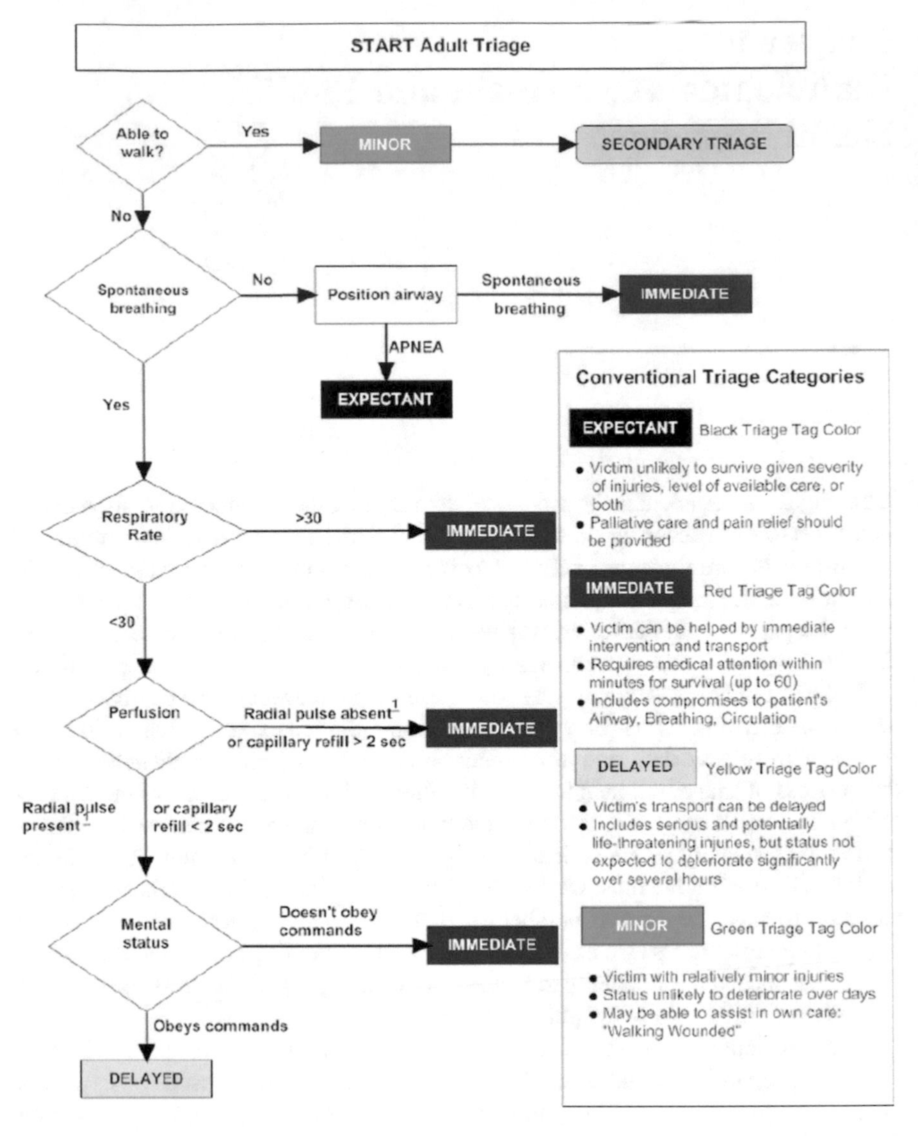

Fig. 18.1 An example of a radiation triage flowchart. Radiation Emergency Medical Management website (https://www.remm.nlm.gov/StartAdultTriageAlgorithm.pdf)

this was chosen to correspond with the lowest level of exposure that can have clinical significance and that should be factored into clinical decision-making.

There are a number of methods for assessing the radiation exposure and/or the amount of radioactive intake a person has received; several are referenced in the flowchart developed by the US Department of Health and Human Services (shown here) and in a special issue of Health Physics devoted to the subject, including a summary by Swartz and several colleagues [14].

18.1 Bioassay

Principle: Bioassay is the process of measuring the amount of radioactivity that has entered the body. There are two major forms of bioassay: in vivo ("in the body") bioassay involves directly counting the amount of radioactivity in the body, typically using scintillation detectors to measure gamma radiation emitted by radionuclides within the body. A chest count, a thyroid count, or whole-body count—holding a radiation detector over the chest or sitting in a whole-body counter—and measuring the count rate emanating from gamma-emitters within the body—are examples of in vivo bioassay. The other, in vitro ("in the glass") bioassay, involves taking samples—generally urine, feces, or blood—and extrapolating the amount of intake from the concentration of radioactivity in the sample.

Procedure: An in vivo bioassay, as noted above, is performed by placing a radiation detector on or near the body near where the radioactivity is expected to be. A person who has inhaled radioactivity that is in an insoluble form, for example, should have the majority of the radioactivity remaining in the lungs; if the person inhaled radioactive iodine then the iodine will enter the blood rapidly and will then collect in the thyroid. Thus, holding the radiation detector close to the body over the lungs or the thyroid (respectively) will produce a count rate; if that count rate is significantly higher than the normal background count rate then we can say that there is radioactivity in the lungs or thyroid. Further, if the counting efficiency of the radiation detector for I-131 is known and the radiation detector has been characterized with respect to the attenuation of I-131 gamma photons as they pass through the tissue overlying the thyroid then we can calculate the amount of radioactivity that has been absorbed by the thyroid as shown in this text box.

Interpreting thyroid bioassay results

Let's assume that a radiation detector has a 10% counting efficiency for I-131 gamma ray photons and 90% of the I-131 gammas emitted from the thyroid are absorbed by the tissues of the neck (for an overall counting efficiency of 1%) then 1000 cps above background would represent 100,000 dps of iodine; since 1 Bq = 1 dps, the thyroid in question would contain 100 kBq of I-131. From Federal Guidance Report Number 11 [16] we can find that 2 MBq (2000 kBq) of I-131 will produce a radiation exposure of 0.5 Sv to the thyroid so we can easily calculate that 100 kBq (0.1 MBq) will produce a thyroid dose of 25 mSv to the thyroid.

In vitro bioassays begin with collecting a sample, typically of excreta. For most radionuclides, especially those that are in a chemical form that is not insoluble, this will be a urine sample, and an aliquot will be extracted from this sample, or the sample will be evaporated to dryness or incinerated to concentrate the radionuclide in order to enhance detection capability. The aliquot will then be counted using an appropriate detector depending on the type of radiation emitted by the nuclide in question and the final activity concentration in the excreta will be used to determine the amount of uptake.

For example, if studies have shown that, say, 1% of ingested radioactivity is excreted in the urine in the first 24 h following an inhalation of a particular nuclide and then we measure 100 Bq of radioactivity in the urine then we can conclude that the person had an intake of 10 kBq of radioactivity, and we can calculate their radiation exposure accordingly (Fig. 18.2).

Sensitivity and accuracy: In vivo bioassays are obtained by counting samples with radiation detection equipment for a given period of time. The sensitivity and accuracy of the analysis depends on a number of factors, including the accuracy of the radiation detector, the amount of time for which the sample is counted, the counting efficiency of the radiation detector for the nuclide(s) in question, and various inaccuracies involved in collecting and measuring the sample. The sensitivity of in vivo bioassay varies from about 1 to 100 Bq and measurement accuracy varies between about ±10–30%, depending on the factors noted above and the nuclide(s) being counted [1].

In vitro bioassays can be more sensitive than in vivo bioassays because there is no tissue providing shielding between the sample and the detector; in vitro bioassays can quantify samples with much less than 1 Bq per sample.

Limitations: This technique requires properly calibrated radiation detectors and counting equipment and the calculated results can vary considerably depending on the person(s) being counted for in vivo bioassay. In addition, every person's body processes radionuclides differently and biokinetic models developed for a reference person likely differ from those exhibited by any individual.

18.2 Lymphocyte Depletion

Principle: Lymphocytes are blood cells that are sensitive to radiation; when exposed to high levels of radiation, lymphocytes begin to accumulate DNA damage and to die. At the same time, the blood-forming organs are also sensitive to radiation so as the lymphocytes die they are not replaced. Since the rate and degree to which the lymphocyte count drops is related to the radiation dose received, measuring the change in the patient's lymphocyte count can be used to estimate the radiation dose to which the patient was exposed (Fig. 18.3).

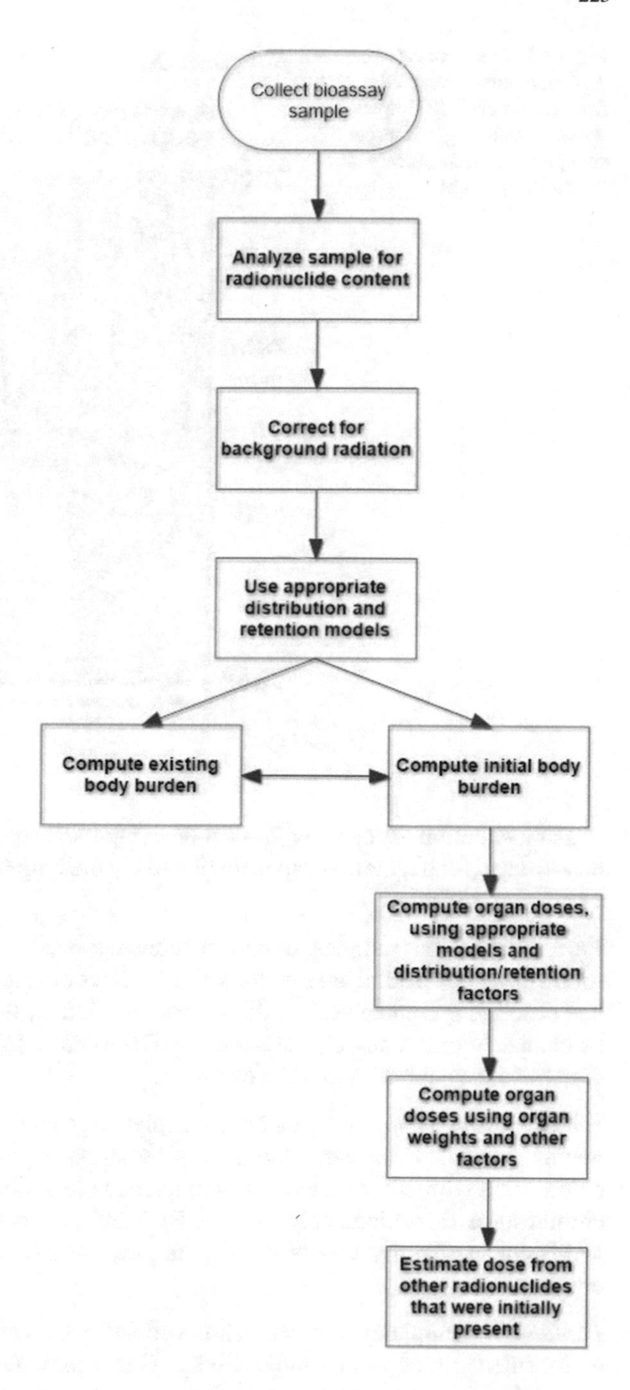

Fig. 18.2 Urine bioassay dose estimation flowchart. Adapted from Boecker et al. [1]

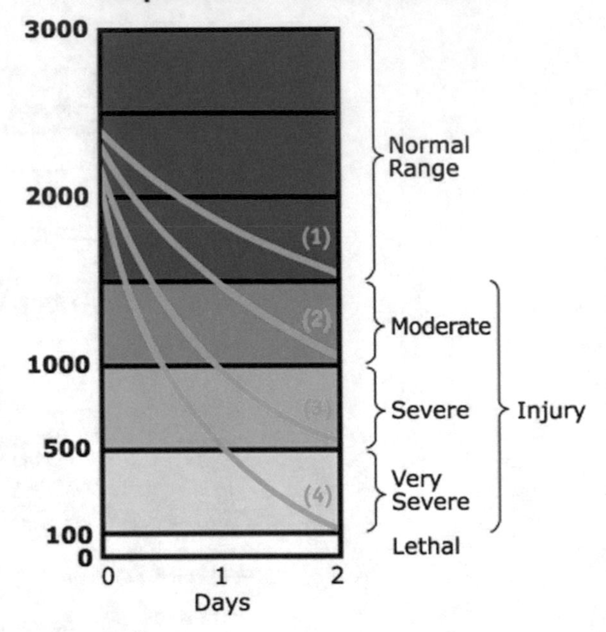

Fig. 18.3 Lymphocyte depletion curve. Adapted from Goans et al. [3] https://www.remm.nlm.gov/andrewslymphocytes.htm. Accessed December 7, 2020

For a radiation exposure of 2–4 Gy the lymphocyte count will drop over a period of 4–6 days; for a radiation exposure of 4–6 Gy the lymphocyte count will drop over a period of 2–4 days.

Procedure: Blood samples should be drawn every six hours for the first week following exposure and are sent for serial CBC (complete blood cell count) to calculate absolute lymphocytes. As the results are plotted, the depletion trajectory can be compared to the models developed by Goans et al. [3] and Guskova et al. [4] to determine the patient's radiation exposure.

Sensitivity and accuracy: Lymphocyte depletion is intended to be a quick and easy method to provide an early estimate of radiation exposure. It should be used in conjunction with other dose estimation methods (e.g. clinical signs and symptoms, chromosome aberration analysis, etc.). By itself, this method is not highly accurate and is not an effective way of determining exposure to doses of radiation less than one or two Sv.

Limitations: Lymphocyte depletion is not effective in cases of partial-body irradiation or for internally deposited radioactivity. This technique is of limited utility if the patient has been exposed to a mixed-radiation field that includes both gamma and neutron radiation.

18.3 Clinical Signs and Symptoms

Principle: Radiation affects the body in a somewhat predictable manner and various levels of radiation exposure cause a predictable suite of symptoms that appear in a predictable sequence. The timing of these symptoms' appearance and their severity can help physicians to determine the radiation dose to which a person was exposed. For example, if a patient begins vomiting within ten minutes of their exposure to radiation then it is likely that the person received a dose in excess of 8 Gy and is unlikely to survive. On the other hand, if the patient has not begun vomiting more than two hours after exposure then they likely received a dose of less than 1 Gy and will almost certainly survive the radiation exposure (Fig. 18.4).

Procedure: Medical caregivers will record the patient's symptoms as they appear and evolve with time post-exposure. These symptoms and the time and order in which they appear are compared to standardized charts or entered into software for evaluation [5, 12].

Sensitivity and accuracy: Doses of less than about 1 Sv cannot be estimated using clinical symptoms as there might be no symptoms other than lymphocyte depletion following lower levels of exposure. At higher levels of exposure the symptoms might not manifest for several hours or days. This method is less accurate than lymphocyte depletion in terms of estimating radiation dose received, given the normal variation between patients.

Limitations: The studies on which these tables were derived generally involved healthy radiation workers and citizens who were exposed to radiation but not to physical trauma. These tables also omit the elderly and the very young, both of whom can be expected to respond differently to various levels of radiation exposure. Thus, the progression of symptoms in patients who were burned, injured, or who inhaled dust and smoke might differ from what is shown in the standard tables [2, 7].

18.4 Cytogenetics

Principle: Radiation exposure can cause chromosomal abnormalities, including dicentric chromosomes, and the number of chromosomal abnormalities is related to the radiation exposure received. The dicentric chromosomes are caused by the misrepair of radiation-induced single-strand DNA breaks; this type of damage is more likely to be caused by radiation than by other agents so this method is considered to be the most specific to radiation exposure. By quantifying the number of dicentric chromosomes in a blood sample it is possible to estimate radiation exposure received [11].

There are additional forms of chromosomal analysis that can be used to determine radiation exposure, including using fluorescent in situ hybridization (FISH)

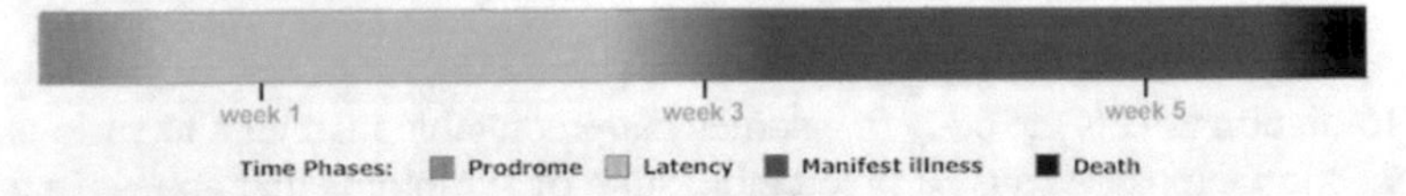

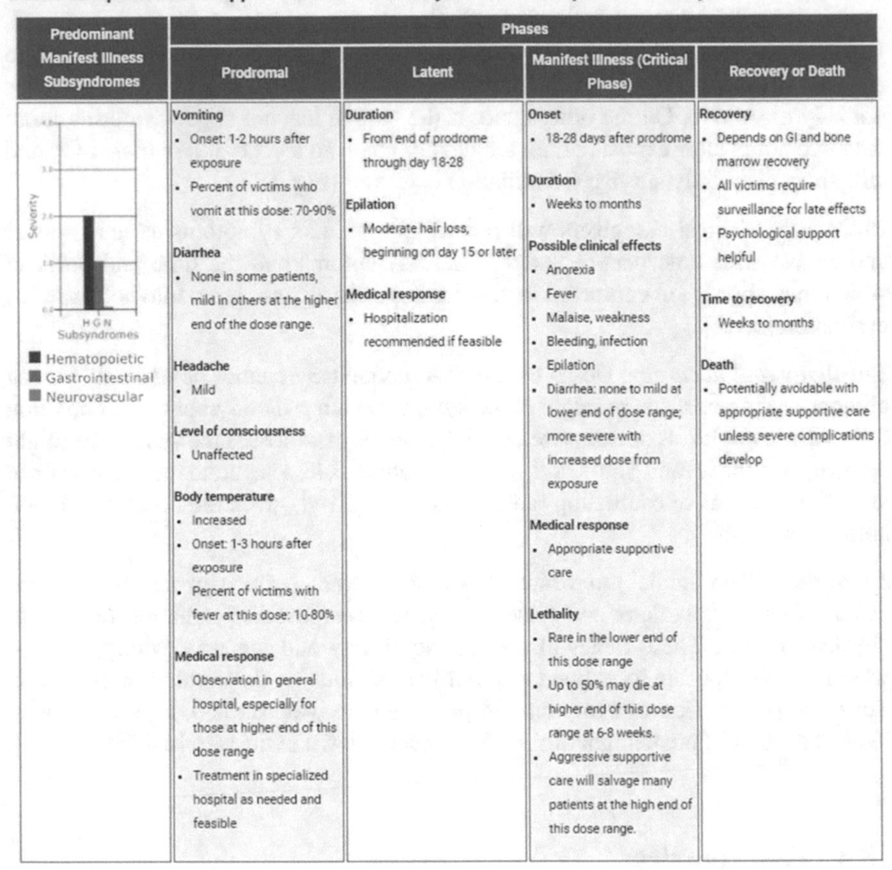

ARS time phases and approximate whole body dose from exposure: 2-4 Gy

Predominant Manifest Illness Subsyndromes	Phases			
	Prodromal	Latent	Manifest Illness (Critical Phase)	Recovery or Death
Severity HGN Subsyndromes Hematopoietic Gastrointestinal Neurovascular	**Vomiting** • Onset: 1-2 hours after exposure • Percent of victims who vomit at this dose: 70-90% **Diarrhea** • None in some patients, mild in others at the higher end of the dose range. **Headache** • Mild **Level of consciousness** • Unaffected **Body temperature** • Increased • Onset: 1-3 hours after exposure • Percent of victims with fever at this dose: 10-80% **Medical response** • Observation in general hospital, especially for those at higher end of this dose range • Treatment in specialized hospital as needed and feasible	**Duration** • From end of prodrome through day 18-28 **Epilation** • Moderate hair loss, beginning on day 15 or later **Medical response** • Hospitalization recommended if feasible	**Onset** • 18-28 days after prodrome **Duration** • Weeks to months **Possible clinical effects** • Anorexia • Fever • Malaise, weakness • Bleeding, infection • Epilation • Diarrhea: none to mild at lower end of dose range; more severe with increased dose from exposure **Medical response** • Appropriate supportive care **Lethality** • Rare in the lower end of this dose range • Up to 50% may die at higher end of this dose range at 6-8 weeks. • Aggressive supportive care will salvage many patients at the high end of this dose range.	**Recovery** • Depends on GI and bone marrow recovery • All victims require surveillance for late effects • Psychological support helpful **Time to recovery** • Weeks to months **Death** • Potentially avoidable with appropriate supportive care unless severe complications develop

Fig. 18.4 ARS symptom timeline. From REMMS [15] (https://www.remm.nlm.gov/ars_timephases2.htm). Accessed December 7, 2020

assessment of radiation-induced chromosomal translocations, premature chromosome condensation assessment, and micro-nucleus assay, and others; each of which has its own benefits and drawbacks [10].

Procedure: A blood sample is obtained 24 h to one month after exposure and is sent to one of a small number of laboratories that are capable of performing this analysis. In the laboratory, lymphocyte growth is stimulated and the metaphase cells are collected, stained, and analyzed for dicentric chromosomes. The number of dicentrics is compared to calibration curves that incorporate the type and energy of the radiation, and is scored. A large variability in the number of dicentrics among scored cells can indicate a partial-body exposure.

Sensitivity and accuracy: Dicentric chromosome analysis is considered to be among the most sensitive method for determining radiation exposure with a lower detection level of about 1 Gy [14].

Limitations: Chromosomal aberration analysis is a complicated process that requires a great deal of skill to perform and to interpret. Determining radiation exposure in this manner can only be done by a handful of laboratories, which might become overwhelmed in the aftermath of a radiological or nuclear attack. In addition, individuals will respond differently to the same dose of radiation so a dose of, say, 2 Gy might cause different numbers of induced chromosomal abnormalities in different patients.

18.5 Quick Assessment of Internal Radioactivity

Principle: In the immediate aftermath of a large radiological event (accident or emergency) there will be a need to quickly assess internal radiation exposure for a large number of people; a Community Reception Center (CRC) that is expected to survey as many as 1000 people per hour might require assessment of internal radioactivity to determine whether the person requires decorporation therapy, requires no immediate attention, or requires further assessment via bioassay. In such a situation there will likely be a large number of people requiring some degree of dose assessment and a very small number of radiation safety professionals available to perform this dose assessment. To help to address this need, Korir and Karam [6] developed a quick assessment methodology utilizing a scoring system that incorporated both qualitative and quantitative factors; designed to be used by personnel with a minimum of training and to be performed in less than ten minutes per person.

Quantitative factors to be evaluated include contamination levels on the face, external bioassay count rate from the lungs, and contamination on the skin; qualitative factors include the presence of dust and debris, injuries, and proximity to the site of the release. In the event of an attack involving alpha or beta-emitting radionuclides the qualitative factors might be the only factors that can be scored.

Procedure: If a person passing through the CRC is found to have elevated count rates they will be sent for decontamination; if the elevated count rates persist then they are sent for assessment, where the person at the assessment station will complete a scoring sheet. Each of the qualitative and quantitative criteria is assigned a score; these scores are summed and the final score determines the person's disposition—a high score warrants presumptive decorporation therapy, a low score warrants later follow-up with a family physician, and an intermediate score calls for further assessment if resources are scarce or for decorporation if resources are plentiful. If the contamination consists of solely alpha- or beta-emitting radioactivity then the qualitative factors and external quantitative factors will still suffice to make this assessment.

The lung count (performed by holding a radiation detector on contact with the chest or back) results are measured in the number of counts above background, which is the quantity that is scored; the score reflects the amount of inhaled radioactivity in the lungs in terms of CDG. In general, a person whose qualitative and quantitative scoring indicates that they appear to have inhaled more than 1 CDG will be sent for decorporation therapy, a person with less than 0.2 CDG will be sent home and advised to contact their personal physician when the opportunity permits, and others will be sent for further assessment using one of the methods described earlier in this chapter.

Sensitivity and accuracy: This method is intended to be a "quick and dirty" screening performed by inexperienced personnel. As such, it is not intended to be highly accurate and it defaults in favor of recommending the assessed for treatment. Since this methodology is intended for use by personnel with minimal training using standard hand-held radiation detectors it is not highly accurate; a high degree of accuracy is not required for a quick screening.

Limitations: Because this methodology is intended to be a quick screening assessment implemented by personnel with minimal training using hand-held radiation detectors it is not intended to be either accurate or precise. However, it compensates for this by making conservative assumptions (i.e. more people will be referred for further assessment or for decorporation therapy who do not require these than those who will be referred to their personal physicians who require further attention).

References

1. Boecker G, Hall R, Inn K, Lawrence J, Ziemer P, Eisele G, Washholz B, Burr W (1991) Current status of bioassay procedures to detect and quantify previous exposures to radioactive materials. Health Phys 60(Supplement 1):45–100
2. British Institute of Radiology (2001) Medical management of radiation accidents: management of the acute radiation syndrome. British Institute of Radiology
3. Goans R, Holloway E, Berger M, Ricks R (1997) Early dose assessment following severe radiation accidents. Health Phys 74(4):513–518
4. Guskova A, Baranov A, Gusev I (2001) Acute radiation sickness: underlying principles and assessment. In: Gusev I, Guskova A, Mettler Jr F (eds) Medical management of radiation accidents, 2nd edn. CRC Press, Boca Raton
5. International Atomic Energy Agency (1998) Diagnosis and treatment of radiation injuries. Safety reports series no. 2. IAEA, Vienna
6. Korir G, Karam P (2018) A novel method for quick assessment of internal and external radiation exposure in the aftermath of a large radiological incident. Health Phys 115(2):235–261
7. Ledney D, Elliott T (2010) Combined injury: factors with potential to impact radiation dose assessments. Health Phys 98(2):145–152
8. National Council on Radiation Protection and Measurements (2008) Report No. 161, Management of persons contaminated with radioactivity, vol 1. NCRP, Bethesda

9. National Council on Radiation Protection and Measurements (2010) Report No. 166: Population monitoring and radionuclide decorporation following a radiological or nuclear incident. NCRP, Bethesda
10. Prasanna P, Moroni M, Pellmar T (2010) Triage dose assessment for partial-body exposure: dicentric analysis. Health Phys 98(2):244–251
11. Ramalho A, Nascimento A, Natarajan A (1988) Dose assessments by cytogenetic analysis in the Goiania (Brazil) radiation accident. Radiat Prot Dosimetry 25(2):97–100
12. Sandgren D, Salter C, Levine I, Ross J, Lillis-Hearne P, Blakely W (2010) Biodosimetry assessment tool (BAT) software—dose prediction algorithms. Health Phys 99(Supplement 3):S171–S183
13. September 11 Victim Compensation Fund (2020) https://www.vcf.gov/. Accessed 23 Nov 2020
14. Swartz H, Flood A, Gougelet R, Rea M, Nicolalde R, Williams B (2010) A critical assessment of biodosimetry methods for large-scale incidents. Health Phys 98(2):95–108
15. US Department of Health and Human Services, Radiation Emergency Medical Management (2020) Radiation exposure: diagnose and manage acute radiation syndrome. https://www.remm.nlm.gov/exposureonly.htm#skip. Accessed 23 Nov 2020
16. US Environmental Protection Agency (US) (1988) Federal guidance report no. 11: limiting values of radionuclide intake and air concentration and dose conversion factors for inhalation, submersion, and ingestion. Government Printing Office, Washington

Chapter 19
Additional Considerations When Responding to a Nuclear Attack

Conventional explosions—even huge explosions such as the 2020 blast in Beirut [7]—pale in comparison to the magnitude of a nuclear explosion. In fact, the relatively primitive and low-yield bomb used to attack Hiroshima was larger than the ten largest man-made non-nuclear explosions combined. Or to look at the matter another way, while a large number of people have a huge amount of personal experience responding to attacks using chemical explosives, there are few—if any—people alive with comparable experience in responding to nuclear attacks. Because of this it seems reasonable to devote some attention to factors that make a nuclear attack so much more devastating than attacks using chemical explosives.

According to the International Atomic Energy Agency [6] the practical goals of an emergency response are to:

1. Regain control of the situation;
2. Prevent or mitigate consequences at the scene;
3. Prevent the occurrence of deterministic health effects in workers and the public;
4. Render first aid and manage the treatment of radiation injuries;
5. Prevent, to the extent practicable, the occurrence of stochastic health effects in the population;
6. Prevent, to the extent practicable, the occurrence of adverse non-radiological effects on individuals and among the population;
7. Protect, to the extent practicable, the environment and property; and
8. Prepare, to the extent practicable, for the resumption of normal social and economic activity.

The first four of these goals will be accomplished in the first few days after the attack and will most likely be accomplished by the city itself, before the national government can provide substantial resources. The next three goals will likely be accomplished in collaboration with the national and (if appropriate) regional govern-

P. A. Karam, *Radiological and Nuclear Terrorism*,
Advanced Sciences and Technologies for Security Applications,
https://doi.org/10.1007/978-3-030-69162-2_19

ment, and the final goal will likely devolve on the local government. The final goal will be discussed in greater detail in the final three chapters of this book.

19.1 The Magnitude of a Nuclear Attack

The primary factor differentiating a nuclear attack from a non-nuclear explosion is the sheer magnitude of the blast and the extent of its effects. For example, the 2020 Beirut explosion killed 190 people and the larger 1917 explosion in Halifax Nova Scotia caused 2000 fatalities, while the 1945 Hiroshima blast killed more than 90,000 and injured tens of thousands. Not only that, but more than 25,000 people died over the first four months following the Hiroshima attack; 25,000 patients who required medical care and resources during the weeks between the attack and their deaths. On top of that, as was mentioned in Chap. 11, there are not enough burn beds in most nations to care for the number of expected burn victims from the thermal pulse and the resulting fires. A nuclear attack will result in a few orders of magnitude more fatalities and injuries than would even the largest conventional explosive attack we have experienced.

There is likely no city in the world that has the personnel, medications, or equipment to care for so many for so long; at the same time, simply leaving so many to die—especially when there are those who might survive with proper care—would be unethical and unconscionable.

All of the other problems associated with a nuclear attack—the radius of destroyed and damaged buildings, the number of destroyed and damaged buildings, the length of the security perimeter, the size of the affected areas—all of the other factors associated with an attack will be larger and more severe than a conventional bomb by at least an order of magnitude if the bomb that goes off is a nuclear bomb.

It also bears mention that a nuclear attack would be the largest event and the most serious emergency that anybody in the city—and likely the national—government has experienced. It is likely that nobody will be able to exercise control over the entire affected area, let alone the entire city, immediately. It is also likely that, when response efforts commence, they will be small-scale and local in nature, gradually coalescing into a more coordinated effort under the control of the Incident Commander and city officials as communications and effective control are re-established.

19.2 Radioactive Fallout

As noted in Chap. 10, a nuclear explosion creates a huge amount of radioactivity and, if detonated as a surface burst this radioactivity will come to ground as fallout in a plume that can stretch 15 km or further downwind of the point of the explosion. Radiation dose rates will be dangerously high in this plume for at least 12 h following the detonation and possibly for as long as a few days. In a large and densely populated city such as New York the fallout plume could affect the health of hundreds of thousands of people with more than a quarter million receiving a potentially fatal radiation dose if they fail to seek shelter following the explosion. There are some considerations that flow from these observations [8].

The first consideration is that hundreds of thousands of lives can be saved if the population understands how to respond between the moment of detonation and about fifteen minutes later when the faullout begins coming down.

- Those who remain outdoors or who seek shelter in a car will likely receive a fatal dose of radiation;
- Those who attempt to flee before knowing where the fallout is being blown are likely to receive a fatal dose of radiation if they are in the fallout plume and do not evacuate in the correct direction;
- Those who seek shelter in a small frame building or close to the outer walls or on the top-most floor will likely receive enough radiation exposure to require medical treatment in order to survive;
- Those who seek shelter in the basement or lower floors of a large building might not even develop radiation sickness.

It is worth noting that the most important decision to be made in the moments after detonation is whether to shelter or flee; this decision will likely be made before the first public service announcement can be broadcast to the public. The population, then, must understand that sheltering is more likely to save their lives than fleeing, and they must know this before an attack takes place. Similarly, parents must understand that they cannot go to bring their children home from school—that they must trust the schools to shelter their children appropriately. Parents who travel to the school to bring their children home—if they are in the fallout plume—will likely be exposed to a fatal radiation dose, as will their children. And both emergency and medical responders must understand that attempting to respond and to help the public immediately might expose them to a fatal dose of radiation. Although the plume will cover only a small fraction of the total area of the city, until its location is known *nobody* in the city should be outdoors.

This means that every person in the city, including emergency responders, medical caregivers, and elected and appointed officials must know that, until the "footprint" of the plume is known, the only safe course of action is to seek shelter immediately. Once the plume outline is known it will be possible to begin planning evacuations for those within the plume, emergency and medical responders who are outside the plume can begin responding to the attack, and both elected and appointed officials can begin marshalling personnel and beginning to plan their response.

The second consideration stems from the size of the fallout plume, which can be several kilometers across and tens of kilometers in length, with dangerously high radiation dose rates extending as far as fifteen kilometers downwind in the event of a surface burst. This is a large area that will be dangerous for as long as one or two days after the time of the attack; the size of the dangerously radioactive area will be far larger than virtually any man-made disaster and will more closely resemble the size of areas affected by natural disasters [1].

The size of the affected area means that the city's emergency response personnel—once it is safe for them to leave their stations—will be stretched thin simply trying to establish and secure the perimeter of the Hot Zone, let alone entering to perform reconnaissance and rescue operations. It will be necessary for those in charge to perform a sort of urban triage; just as medical personnel cannot spend time attending to those beyond saving and should not spend time on those whose lives are not at risk, neither should limited city resources be concentrated in areas that are beyond saving (e.g. the rubble zone and within the area affected by the mass fire) or in areas in which there is little that needs done and there are few or no lives to be saved.

The third consideration is that the numbers of affected people are likely to overwhelm available resources for the first weeks or even months after the attack. Consider: there can be nearly one million people evacuating from the plume footprint and another half million or more from the areas adjacent to the rubble zone. In the United States, an attack against a major city such as New York City, Chicago, or Los Angeles can produce as many eventual refugees as the total populations of all but a handful of American cities; an attack against Karachi, Seoul, London, Paris, Moscow, or any of the major cities in India and China would produce as many, or more. The scale of the medical and humanitarian needs in the affected city might initially overwhelm the nation's available resources. Caring for those evacuated and displaced, as well as those whose homes remain safe but who lack electricity, food, and/or water will be challenging and may prove to be beyond the capabilities of the majority of nations, just as the sheer number of those exposed to high levels of radiation and radioactive fallout is likely to overwhelm the resources available to perform even a cursory radiological dose assessment.

19.3 Lack of Public Preparation for Civil Defense

During the Cold War, the US, USSR, and many or their allied nations devoted significant resources towards civil defense. Most Americans who were born during the twentieth century were exposed to civil defense advertisements, public service announcements, signs, and more; it is safe to say that the majority of Americans who lived in the twentieth century had heard the phrase "duck and cover" and had participated in drills about what to do in order to try to save their lives in the event of a nuclear attack. The same was true in the Soviet Union, as well as in most of the nations of NATO and the Warsaw Pact.

Since the end of the Cold War, much of this has changed. The majority of Americans born after 1990 are likely to have never seen (or to not remember) nor to have received any civil defense training whatsoever. The fact is that the majority of Americans, and the majority of citizens in nations throughout the world, have had little or no training in civil defense. As a result, it is possible that many people in a city that was attacked—as many as tens or even hundreds of thousands—do not know what actions to take to keep themselves safe. As a result, hundreds of thousands of people might die simply because they have never been told how to save their lives in the immediate aftermath of a nuclear attack.

Part of the reason for the recent lack of civil defense training in the US is a reluctance to risk frightening the public. The author participated in multiple telephone conference calls involving local, state, and Federal government officials in which some of these officials stated that, while they understand that the public could not expect to receive civil defense instructions in time to be helpful following a nuclear attack, they nevertheless advised against providing such information before an attack occured for fear of frightening the public, under the assumption that the public will assume any such instructions might indicate secret governmental knowledge that a nuclear attack was imminent. Because of this, the American government has been, as of this writing, unwilling to provide advance information on how to survive a nuclear attack. Instead, the plan is to provide a public service announcement immediately following an attack, assuming that government can be assembled and that radios and transmission equipment are operational. For reasons outlined below, these might not be good assumptions.

19.4 Continuity of Government

In the immediate aftermath of a nuclear attack it might not be evident which elected and appointed officials are able to communicate, or are even alive. It might well require several hours to determine if various officials have survived the attack, if they are trapped by high levels of fallout, are injured, or have otherwise been rendered incapable of participating meaningfully in governmental decision-making. Not only that, but those officials who have not been killed by the attack might be unable to travel because of debris and/or high radiation levels and might be unable to even communicate due to EMP-inflicted damage. Thus, it might not be possible to quickly assemble a working government capable of making decisions and managing the emergency response.

Assembling the government might take attention away from the initial response efforts, especially if communications and travel are difficult (or impossible). But doing so is vitally important as those who head the major agencies and senior elected officials are the ones with the broad view of the city as a whole and who are familiar with attending to the needs of the city as a whole; these are the people who can help knit together the efforts of the disparate agencies into a coherent response. In addition, working at a slight remove from the scene, it is these people who are able to keep in mind the longer-term goals and to help coordinate the various agencies' response efforts to remain on a path towards those goals. For this reason, it is essential to establish some form of government as quickly as possible after the attack takes place.

This means that the city should develop multiple plans for contacting the core governmental officials to determine who has survived, their location, and to bring them together in person or by some other means (telephone, radio, etc.) so that they can begin the task of overseeing the emergency response while keeping the city running, asking for assistance, and the other work of government. "Multiple plans" should also include contingency plans for assembling lower-level officials in the event more senior personnel cannot be contacted.

Having plans is a good start, but these plans must be communicated to those they affect and they must be practiced as well; the Mayor (for example) needs to know who should be called to let City Hall know that they are alive, and the person taking that call needs to know who to try to contact if the Commissioner of Health fails to establish contact within 30 min (or whatever time frame is agreed upon). And all of the senior officials should know how many of them need to assemble to form a quorum authorized to make decisions on behalf of the city. Until such a plan exists, until there is a list of names and several alternates for each, and until such a plan has been practiced by all involved, the city cannot be said to have an assured continuity of government and might not be able to provide a coherent and effective response in the immediate aftermath of a nuclear attack.

19.5 Electromagnetic Pulse (EMP)

When a nuclear weapon explodes it sends an intense pulse of ionizing radiation into the environment. This ionizing radiation creates pairs of charged ions in the atmosphere; the light electrons can travel further and faster than the heavier ions, creating a separation of electrical charge in the atmosphere. This separation of charge, in turn, gives rise to an electrical current as the positive and negative charges flow back towards each other to recombine. It is this process that drives the formation of an electromagnetic pulse that, in turn, can induce electrical currents in pipes, cables, and other conductive materials [4].

The effects of EMP were hypothesized, but were not fully appreciated until 1962, when a high-altitude atmospheric test called Starfish Prime caused large-scale electrical problems in Hawaii, over 1400 km away [3]. However, due to the ban on atmospheric nuclear weapons testing, only a handful of tests could be studied for their EMP effects. Thus, EMP theory is more complete than are observations of EMP.

However, we do know that the currents induced by EMP can cause fires, trip circuit breakers, and can damage or destroy unshielded electronics. It also appears that the effects of EMP are limited in radius to the line-of-sight distance to the horizon from the location of the detonation; for a surface burst, this is about 10 km or less; for a high-altitude burst, this can be more than 1000 km. Within that radius, we should expect that the EMP will damage or destroy all computers, communications equipment, and other electrical and electronic gear that was operating at the time of the EMP. It is also reasonable to assume that parts of the city's electrical generation and distribution system will be damaged or rendered inoperable; if the radius of the EMP includes transformer yards and power generation facilities then electrical power can be lost over large swathes of the city.

Consider a city attacked with a nuclear weapon. In many cities, large portions of the subway tunnels are beneath the local water table; in cities that are located on rivers, lakes, or the ocean these tunnels might be lower than sea (or lake or river) level. The tunnels are kept dry by de-watering, and if the dewatering pumps lose their source of power then the tunnels will begin slowly filling with water. If not reversed, this water will eventually rise to the level of the local water table, flooding the tunnels. If members of the public sought shelter in the train stations, they might be trapped by the rising waters.

Meanwhile, the Mayor and their deputies, the heads of the Police, Fire, Health, and Emergency Management agencies are unable to call in because their telephones and radios are not working, the Mayor and Health Commissioner are unable to make public service announcement instructing the population to seek shelter, the public is unable to check websites for information and instructions, the Police and Fire Commissioners are unable to remind their personnel to shelter in their stations until the plume can be mapped, and what response efforts that exist are likely to be disjointed and uncoordinated.

As the city begins to recover from the EMP effects it will be possible to begin to be possible to establish communications with emergency responders, transmit public service announcements, establish the Incident Command and Emergency Operations Center, and to restore electrical power. If communications and power are restored relatively quickly the impact on emergency response activities and public health will be minor; if they are not restored for several hours or longer then it will likely hinder response efforts.

This being said, it is also possible that the impact of EMP will not be as severe as imagined. Computers, radios, electronics, and electrical systems of today are different from those that existed a half-century ago and, while many of them have been tested, none have been exposed to the EMP from an actual nuclear weapon. Thus, it is possible that EMP will have only a moderate impact on the emergency response. The point is that we do not know; that being the case it makes sense to assume that EMP will knock out or disable at least some systems and to plan on how to maintain operations in spite of that.

19.6 Logistics

The author participated in a multi-day nuclear terrorism exercise that postulated an attack near, but not in New York City. As the exercise progressed, only a small part of Manhattan was affected by fallout and no part of the city was directly affected by the blast. In spite of this, as the exercise played out, the entire city began to experience what can only be called "logistical effects" resulting from the explosion. In particular, the high radiation dose rates and contamination levels from the plume were sufficient to shut down the major roads, bridges, and tunnels leading into Manhattan from New Jersey and southern New York. Without truck traffic, food shipments into New York City were forced to cross Staten Island and the Verrazano Narrows Bridge, then to wend their way through Brooklyn to reach Manhattan and Queens. The loss of access to the George Washington Bridge, the Triboro Bridge, and the Lincoln and Holland tunnels, all of the food for more than eight million people had to travel along a single corridor, one that was also being used by emergency response vehicles, government vehicles, and personal vehicles.

New York City is a major port so the exercise looked at bringing food in by sea, only to realize that the majority of the docks were on the New Jersey side of the harbor and were also cut off from the city by the fallout plume. Air cargo was also investigated; one airport was also located in New Jersey and was unusable, but two other airports were available. The problem is that airplanes do not carry as much volume as do ships and are more a more expensive mode of shipping than trucks—the city's airports could help, but were not sufficient to meet the city's needs either.

What quickly became clear to the participants was that a nuclear weapon detonated in New Jersey could affect more people in New York City than in New Jersey. By cutting the city off from the harbor and all but one road, a single (simulated) nuclear weapon with a relatively low yield was able to affect food and other supplies to more than eight million people. In many parts of the world such a temporary interruption in supplies would be an inconvenience; in some cities, however, the typical resident might keep only a few days' worth of food in their home, relying on frequent shopping or on food delivery from local restaurants; in such cities, even a small interruption in the transportation of food and other supplies can put lives at risk. As an example of this, Figure 19.1 shows the limited number of access points into New York City.

As a result of this exercise, one conclusion was that the city should take immediate action to transition the airports immediately from passenger to cargo facilities, realizing that New Yorkers needed to eat and that it was also essential to be able to bring in the personnel and equipment to begin to address the emergency. Finding docks for cargo ships was also important if New York City might be cut off from the majority of port facilities. Finding ways to reopen ground transportation routes was also essential, along with convincing drivers that these routes were safe to use.

This exercise made it clear that logistics can be a major problem in the immediate aftermath of a nuclear attack. But this had become clear to the author several years earlier, during a trip to Japan a month after the earthquake, tsunami, and nuclear reactor meltdowns in Fukushima. Over the course of discussions with officials in various cities in the area affected by the accident the team heard that, even one month after the accident, the logistics of continuing to feed, clothe, and provide supplies for residents, responders, and workers in the affected area continued to be an issue. Considering the relatively low population of that area compared to many major cities (hundreds of thousands versus millions), this highlights the importance of ensuring that supply routes are opened quickly and that they remain open.

Fig. 19.1 The New York City area, showing potential air, land, and sea resupply routes. There are other possibilities (e.g. further out on Long Island), but these are hampered by distance, lack of high-volume port facilities, lack of primary roads, and so forth. The biggest hinderance is the relatively small number of ground routes into the city; the loss of even a single bridge or tunnel would significantly reduce the transportation of supplies into Manhattan, Queens, and Brooklyn

19.7 Sheltering and Evacuation

At some point it will be necessary to evacuate the public from areas that are not safe. Even a person sheltering in a safe location within the fallout plume can receive enough radiation to cause radiation sickness over the course of several hours, and that can become life-threatening after a few days.

There are risks other than radiation exposure of course. Those who are sheltering in areas affected by the blast and thermal effects might be injured and, even if their injuries are not life-threatening (e.g. broken bones), they will need to receive medical care. Others might have only limited food and water (especially if power is lost to refrigerators and municipal water pumps), still more might have lost refrigeration for their medications or might need to refill prescriptions. In extreme weather there will also be the need to get those who have lost heating or air conditioning to safe locations.

So, for a number of reasons both radiological and non-radiological it might be necessary to evacuate several hundred thousand people from the area. If the city is also affected by other issues—loss of water, prolonged loss of electrical power, inability to bring in food and supplies, and so forth—then it might be necessary to evacuate several million people in the first few weeks following an attack.

This means that there are multiple populations that might require evacuation, with overlapping needs and different time constraints.

Those living in close proximity to the site of the explosion are most likely to be suffering from injuries, to have lost utilities, to be in damaged buildings, to be freezing (or sweltering) in extreme weather, to be unable to fill their prescriptions, and so forth. They are also likely to have received the highest radiation exposure from the explosion. For those who are outside of the fallout plume (up to 75%) the radiation exposure will end less than one second after the detonation; for those who are also downwind of the detonation the exposure will continue until they are evacuated.

Those living more than a few to several kilometers from the scene of the explosion are likely to be outside of the areas affected by blast, fire, and prompt radiation from the explosion; unless they are within the plume's footprint they are not likely to be directly or immediately affected by the attack, although they can certainly be affected by the logistical issues discussed earlier as well as EMP effects in the areas closest to the explosion. In fact, for those who live, say, fifteen kilometers from the explosion and who are not within the footprint of the plume, there might be little to no immediate impact, and possibly even little impact as time goes on—while just a few kilometers away radiation dose rates in the plume might be dangerously high.

Evacuating people from non-radiological events is common; humanity has become skilled at evacuating people following any number of natural disasters as well as terrorist attacks, urban disasters (e.g. building collapse, steam line ruptures, etc.), and so forth. Evacuating people from nuclear disasters (e.g. Chernobyl, Fukushima) is very uncommon, but it is a process that has been well thought-out.

In 2014, Michael Dillon, a scientist at Lawrence Livermore National Laboratory published an interesting study examining the optimal amount of time to shelter following a nuclear detonation [2]. As a part of his study Dillon summarized eight previous studies in this area, noting factors affecting the optimal sheltering time, the optimal sheltering time for those in poor shelters, and the optimal sheltering time for those in good shelters. Factors Dillon identified as affecting sheltering time in these various papers included:

- Shelter quality (the protection factor provided by the shelter)
- Knowledge of fallout pattern
- Delay until evacuation is possible
- Distance from detonation
- Outdoor radiation dose rate
- Other hazards (e.g. fire)

- Medical needs
- Food and water
- Operational and logistical considerations
- Indoor radiation dose rate
- Likely radiation dose during evacuation
- Number of people to be evacuated
- Evacuation method, route, and starting location
- Specific response strategy.

In these various papers, the optimal sheltering time varied from a few hours to several days with most recommending a sheltering time of about one day.

The radiation exposure received by those being evacuated depends on:

- External radiation dose rate
- Protection factor provided by the structure in which one is sheltering (for example, those shown in the above graphic)
- The length of time spent outdoors during evacuation.

For example, a person about 3 km from a 10 kt nuclear explosion who tries to evacuate immediately—as the fallout is still settling to the ground—can receive a dose of 5–7 Sv of radiation exposure, putting them at risk of death [1]. On the other hand, if that same person spends 10 h sheltering in a building with a protection factor of 10 (as shown in Fig. 19.2) and then evacuates over the same route their total

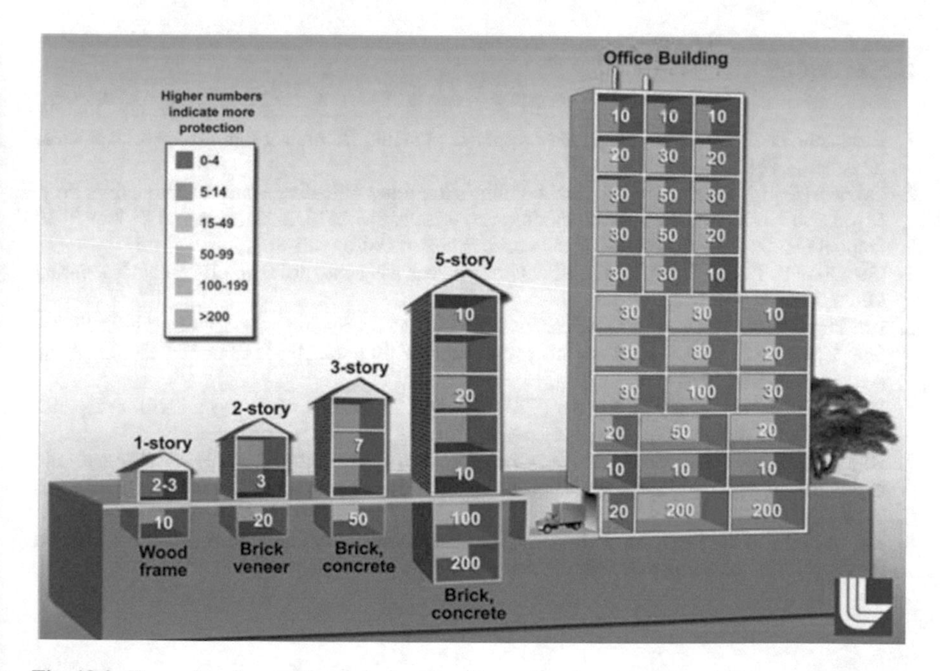

Fig. 19.2 Examples of protective factors offered by various locations in several types of buildings. LLNL [1]

exposure will be only about 1.5 Sv, causing mild radiation sickness but no more. The optimum time to evacuate—the time that will produce the lowest total dose—will vary from place to place but, in general, the dose curve flattens out considerably after about three or four hours so the difference in total dose from evacuating at, say, 12 or 24 h is not substantially different than the dose from evacuating after 4 or 6 h. This assumes that a safe and quick evacuation route has been identified.

Another factor to consider is the overall safety of those being evacuated. We have already discussed the hazards from unstable buildings, falling glass, ruptured utilities, and so forth; to these we must add the potential for civic unrest as a panicked populace and/or opportunistic criminals take to the streets. The goal should be to reduce, not just the radiation dose, but, rather, the *total risk* to which the affected members of the public are exposed. In some circumstances it might be reasonable to wait for an evacuation route to be secured, even if it takes extra hours for the route to be secured.

It must also be recognized that evacuation is not necessarily a benign process. Over 2200 people died during the evacuations following the Fukushima accident, most of whom would have survived had they remained in place [5]. This is not a criticism of the decision to evacuate—the government of Japan followed existing guidelines and established best practices when the evacuation was ordered. It is only in retrospect, knowing the amount of radioactivity that was ultimately released and the resulting contamination levels, that this became known. But the fact remains that evacuations can cost lives.

References

1. Bruddemeier B, Dillon M (2009) Key response planning factors for the aftermath of nuclear terrorism, LLNL-TR-410067. LLNL
2. Dillon M (2014) Determining optimal fallout shelter times following a nuclear detonation. Proc. Royal Soc A 470:20130693. http://dx.doi.org/10.1098/rspa.2013.0693. Accessed 24 Nov 2020
3. Dupont D (2004) Nuclear explosions in orbit. Sci Am 290(6):100–107
4. Glasstone S, Dolan P (1977) The effects of nuclear weapons, 3rd edn. Government Printing Office, Washington DC
5. Harding R (2018) Fukushima nuclear disaster: did the evacuation raise the death toll? Financial Times March 10. https://www.ft.com/content/000f864e-22ba-11e8-add1-0e8958b189ea. Accessed 24 Nov 2020
6. International Atomic Energy Agency (2003) Method for developing arrangements for response to a nuclear or radiological emergency. IAEA, Vienna
7. Rigby S, Lodge T, Alotaibi S, Barr A, Clarke S, Langdon G, Tyas A (2020) Preliminary yield estimation of the 2020 Beirut explosion using video footage from social media. Shock Waves 30:671–675
8. US Department of Homeland Security (2016) Quick reference guide: radiation risk information for responders following a nuclear detonation. US DHS, Washington DC

Chapter 20
Remediation

When the smoke has settled and the dust has cleared—when the emergency phase has ended—attention will turn to restoring access to the affected areas. This will almost certainly include remediating the contaminated areas (unless the nuclide(s) used are short-lived), but the remediation is not the first step; before remediation and cleanup can begin, there must be some agreed-upon basis for the remediation, as well as an agreement as to whether or not the same standards should be used everywhere (homes, parks, parking lots, and so forth). In the US, the Federal government has developed a bewildering welter of regulatory requirements that are relevant to managing the aftermath of a radiological attack [2]. However, it has (as of this writing) declined to recommend cleanup standards for the aftermath of a radiological or nuclear attack in order to let the community(s) attacked choose standards that they feel are appropriate; this might delay the eventual restoration of access and reoccupancy of the affected areas. In the short term, the final cleanup standards might also affect the decontamination techniques used as well as decisions to decontaminate versus discard objects (or entire buildings) that are simply not worth the time and cost of decontamination. This, in turn, can have an impact on the amount of remediation waste generated as well as the eventual cost of remediation and waste disposal.

20.1 Who Determines Remediation Standards?

The most fundamental question that must be answered when performing a decontamination is "When are we done?" Of course, this most fundamental question cannot be answered until the "we"—the people answering this question—can be chosen.

One potential solution to this problem is for the national government to develop remediation standards, aided (of course) by advisory committees comprised of

P. A. Karam, *Radiological and Nuclear Terrorism*,
Advanced Sciences and Technologies for Security Applications,
https://doi.org/10.1007/978-3-030-69162-2_20

esteemed professionals. Such standards would have the advantages of being consistent across the nation, of being developed by highly competent professionals who almost certainly understand the science well enough to have a sound basis for the limits they derive. In addition, by developing cleanup standards in advance there would be ample time to publish drafts to solicit comments and feedback from both the public and the scientific communities and they would be available to be used at the time of an attack, rather than requiring weeks or months to develop in the aftermath. Unfortunately, having so important a decision made by strangers who do not have a stake in the eventual outcome might not be acceptable to the citizens since many of those developing remediation recommendations will come from outside the community. In addition, government "blue ribbon" committees are not necessarily trusted to have the best interests of the local citizens at heart.

Another potential solution is to wait until there is an attack and to then assemble a panel of "stakeholders" (those who have a stake in the matter) to develop cleanup standards in the aftermath of an attack. In a case such as this the national government might develop guidance documents in advance, but the final standards would be developed by those living in the city that was attacked and who have ties to the community. Such a decision would be more likely to be respected by residents, at the same time, it is also more likely to be driven by the concerns of people who might not have a good understanding of radiation science.

Regardless as to how decontamination criteria are established, they should have a basis. Risk-based standards, for example, are based on the risk posed to the population by radiation dose rates, internal contamination due to resuspended radioactivity, and any number of assumptions to help determine how significant these exposures might be. These assumptions would require discussion, justification, and acceptance by the group developing cleanup standards, and would need to be acceptable to the public as well.

Whatever standards are decided upon, if they are to be accepted by the community then those who developed them must be accepted as not only having appropriate expertise, but also as representing the interests and concerns of the various stakeholders: the public, business, public health, government, and whichever other groups are represented. If those developing cleanup standards are not accepted by the various stakeholders then it is entirely possible that the standards will also be rejected. This means that the group determining remediation and reoccupancy standards must be seen as scientifically competent, but also as being attuned to the interests and needs of the various groups that make up the community stakeholders.

20.2 Contamination Limits and Decontamination Goals

Many decontamination and remediation projects are driven by contamination levels, often based on counting statistics and detector characteristics to determine when contamination is unambiguously present. However, statistics-based contamination limits have very little to do with health and safety; using the dose conversion factors from Federal Guidance Report #15 [14] we can easily calculate the radiation dose rate resulting from any level of contamination from any of a large number of contaminating radionuclides. With Co-60, for example, a surface contamination level of 10 dpm cm^{-2} (a common cleanup limit in the US) will yield a dose rate somewhat lower than 0.020 μSv hr^{-1} and will produce an annual exposure of only about 175 μSv yr^{-1} to a person spending 100% of their time on this surface. Living on such a surface for 80 years would produce a lifetime dose of about 14 mSv, comparable to the dose from a single whole-body CT scan. Clearly this is not a significant health risk. Remediating to such low contamination levels adds very little risk reduction at a great cost, especially when we consider that how unlikely it is that any single person would spend 100% of their lives in so restricted an area.

Setting cleanup limits based on the lifetime risk of developing a fatal cancer is another possibility. With a risk coefficient of 5% added fatal cancer risk Sv^{-1} we can calculate the dose rate and corresponding contamination levels to produce a given level of risk, as shown in this text box.

When confronted with contamination it is not uncommon for the public to demand that it be cleaned up to pre-accident background levels. This is understandable, but it offers little actual risk reduction at a potentially high cost. From an "aesthetic" standpoint, this standard would return the affected areas to more or less exactly as they had been prior to the attack; from a health and safety standpoint it makes little sense.

It is also possible to remediate to a level based on risk—the total radiation exposure—to the general population, as described in the following text box. This is the most cost-conscious cleanup standard, but this would also leave behind the highest levels of contamination and might prove to be unacceptable to members of the public who are inordinately frightened of radiation and radioactivity [1].

Calculating a risk-based contamination limit

Assume that a government or advisory committee has determined that they will remediate an area until the risk posed by that area is no more than 0.01% (one fatal cancer for every 10,000 people) to a person spending 100% of their time on this site over an 80-year lifetime. Having determined the acceptable risk, it is necessary to determine the level of contamination that will generate this level of risk.

Using a risk factor of 5% Sv^{-1} we can easily calculate the lifetime dose that will produce a risk of 10^{-4}; dividing 10^{-4} by 0.05 Sv^{-1} shows us that a lifetime radiation dose of 0.0002 Sv (0.2 mSv) will produce this level of risk. Over an 80-year lifetime, this is equivalent to 2.5×10^{-5} Sv yr^{-1} or slightly less than **3 nSv hr^{-1}** since a year has 8760 h.

According the Federal Guidance Report #15 [14] the radiation exposure from Cs-137 contamination is 7.85×10^{-18} Sv s^{-1} (or 2.83×10^{-14} Sv hr^{-1}) for every Bq m^{-2} of contamination. Using this conversion factor we can calculate that, to produce a dose rate of 3 nSv hr^{-1} requires a surface contamination level of about 10 Bq cm^2. The equates to 1000 Bq for every 100 cm^2, which corresponds to 60,000 dpm for every 100 cm^2; this is significantly higher than the 5000 dpm limit for total contamination that is usually used as a limit in the US. And, while causing very low risk, this level of contamination, and the corresponding count rate when surveying, would likely be alarming to members of the public.

In general radiation dose rate—and long-term dose and risk—is directly proportional to the contamination levels in the area in which one is living. If contamination levels are increased by a factor of, say, five then the dose rate also increases by a factor of five, the total dose increases by a factor of five, and the risk (assuming that risk is directly proportional to dose at all levels of exposure) increases by a factor of five.

So now consider a hypothetical contaminated area in which contamination levels drop linearly with distance, in all directions. Since the volume of a sphere (or, in this case, a half-sphere) is proportional to the radius cubed, the volume to be remediated in this scenario will increase with the cube of the degree of remediation to be accomplished. In other words, reducing the contamination levels (and dose) by a factor of two will require excavating eight times as much soil with a corresponding increase in cost. Thus, reducing risk by a factor of two increases the expense of remediation by a factor of as much as eight [4]. This is the fundamental tension when developing cleanup levels—the ever-increasing cost of further remediation that must be balanced by the ever-diminishing reduction in risk "purchased" with these expenditures.

Of course, cost is not the only trade-off. Excavating contaminated soil, demolishing contaminated buildings, loading contaminated materials into containers and shipping those containers hundreds or thousands of kilometers, then unloading and burying them at the other end … all of these carry their own risks and at some point the added risks these activities generate might well be greater than the risks they help to avert.

To the remediation and transportation risks we must add the risks that arise simply from spending the public's money. There have been a number of studies (e.g.Keeney [5]) that have concluded that large costs distributed across society can have a negative effect because, in order to pay higher taxes (or utility bills or other necessary expenses) the public has to take money from other risk-reduction measures. They might delay replacing worn tires, be unable to pay for medications, choose not to see a physician, and so forth. Keeney noted a rate of one anticipated fatality for about $25 million in cost distributed across society; thus, a billion-dollar remediation might lead to as many as 40 premature deaths. If the contamination that was remediated produced only a very small radiation dose, simply funding the remediation might cost more lives than the contamination would have.

We have already discussed the tension between risk reduction and remediation costs; another tension in developing remediation limits is that of scientifically justifiable risk-based limits versus the perception of those who do not have a strong understanding of the science and calculations underlying the risk calculations.

Thus, a citizen with a radiation detector might be alarmed at the count rate they are measuring, even if the dose rate is too low to pose a great risk. Alarmed by seeing high count rates, it would be natural to demand cleaning up until the count rate drops to normal background levels measured elsewhere in the city. At the same time, radiation safety experts would note the high cost of remediation to those levels compared to the marginal reduction in risk.

Comparison risk-based and technology-based remediation standards

Using the same sort of calculations described in the earlier text box we can calculate that a Co-60 contamination level of 1000 dpm per 100 cm^2 (0.17 Bq cm^2) will produce a radiation dose rate of about 150 nSv hr^{-1} for an annual radiation exposure of about 1.3 mSv yr^{-1}. A person living on this surface for 80 years would receive an integrated exposure of about 10 mSv; using the linear no-threshold hypothesis and a slope factor of 5% risk of developing a fatal cancer for 1 Sv of exposure this person would have a lifetime cancer fatality risk of 0.005%, or about five for every 100,000 residents. This risk can be adjusted to account for spending part of one's time in areas with more or less contamination.

Assume that the city's Radiological Advisory Committee (RAC) agrees that an acceptable risk level is one possible cancer fatality for every 100,000 residents (which would be stated as a cancer fatality risk of 1×10^{-5}) then the acceptable average contamination level would be 0.035 Bq cm^{-1}.

While this level of contamination produces very little risk, it will produce an elevated count rate, although the actual count rate will depend on the instrument being used. A small scintillation detector with a cesium iodide detector would have a count rate on the order of several thousand counts per minute while a GM tube might register only a few hundred counts per minute (although even this is much higher than the normal background count rate of 50–100 CPM). Thus, as with the previous case, an area that poses very little risk might still produce a sufficiently high count rate as to alarm the public.

20.3 Decontamination Standards for Homes, Businesses, Public Spaces, etc.

Another question that must be answered is whether or not to apply the same decontamination standards universally or to adjust them to account for the fact that different areas have different uses and occupancy factors. There are 168 h in a week; consider a person who spends those hours thusly:

- Work 40 h
- Home 100 h
- Commuting 10 h
- Shopping, errands 8 h
- Exercise 10 h.

Now, consider the questions that even this simple list raises:

- Should roads and train tracks have the same cleanup levels as home?
- Should parks, public squares, and related areas have the same cleanup levels as the workplace (to protect those who work to maintain them) or as the roads (to reflect the way the typical person uses them)?
- Should alleys have the same cleanup levels as main streets—and should sidewalks have the same standards as the roadways?
- Should the nature of the surface in each location be considered since they affect the amount of contamination that is resuspended and might be inhaled?

Determining what should constitute allowable levels of surface contamination can become hopelessly complex if every type of surface and the use patterns of every area is taken into account. It might be easier to try to weight all of these various factors to develop a single cleanup level to use everywhere, accepting the fact that a large number of residents are likely to spend their time differently than the averages assumed for the purpose of calculation. As a result, this approach would require applying a safety factor to account for, for example, stay-at-home parents, workaholics, nature enthusiasts, train crews, highway workers, and so forth.

Radiation dose rate from contamination on the ground (or floor) is important, but equally important (possibly even more important) is determining radiation dose rate from the intake of radioactivity, which is largely the result of resuspended radioactive contamination. This is especially important when the nuclide(s) used are alpha-emitting actinides such as plutonium or americium, or bone-seeking elements such as strontium or radium. Unfortunately, determining resuspension factors is fiendishly difficult because it depends on the physical and chemical properties of the contaminant, the physical and chemical properties of every surface (asphalt, concrete, brick, grass, soil, gravel, marble (both polished and varying levels of unpolished), glass, etc.), the variability in wind speeds (as well as breezes from traffic passing at various speeds), the orientation of surfaces (horizontal, vertical, various inclines), chemical reactions between the contamination and the various materials that comprise all of the different surfaces, and much more. For this reason, determining resuspension factors and the resultant intake is a problem that, at present, cannot be solved in any meaningful manner. Accordingly, if internal exposure from resuspended radioactivity is to be considered in developing decontamination standards, it will be necessary to make a number of simplifying assumptions; such determinations might be better performed in advance of the need for this information.

Another factor to consider is how much remediation is required. Consider, for example, a 40-story office building with extensive contamination on the exterior surfaces. How much of the exterior should be decontaminated or replaced?

It is reasonable to decontaminate or replace exterior surfaces on the ground floor because people can be exposed to radiation from embedded or affixed contamination. It might make sense to decontaminate the second story as well for the same reason. But should the 20th story be decontaminated if dose rates on the building's interior are acceptable? At a height of 60 or 70 m above the ground the contamination would produce little or no dose at ground level; does it make sense to spend millions—even tens or hundreds of millions—of dollars to decontaminate or replace contaminated windows and exterior walls that produce little or no risk? Remediation or decontamination can be expensive and might produce little or no risk reduction; or will the contamination pose a sufficient risk to window-washers, or to members of the public as rain and weathering brings it down to street level, to justify the expense of cleanup or removal? There are circumstances in which it might be most sensible to simply leave the contamination in place rather than spending vast sums of money on remediation that will accomplish little or no actual reduction in risk (for example, a 2019 contamination accident at the University of Washington is estimated to have a remediation cost in excess of $60 million—Leone [6]).

There are national and international recommendations regarding the acceptable dose to those living in an area to help determine when an area should be evacuated and when the population should be permitted to return; in the US these are referred to as Protective Action Guides (PAGs). The US Environmental Protection Agency (EPA), for example, recommends evacuating from areas where even those sheltering in place are expected to receive a dose in excess of 10 mSv during the first four days if evacuation will produce a lower dose than sheltering. The EPA further recommends relocating members of the public if the projected dose in the first year will be 20 mSv with 5 mSv in any subsequent year [13]. At the same time, a 2014 NCRP report notes "The RDD/IND Planning Guidance (from DHS [11]) does not include specific cleanup criteria for RDD or IND incidents. Rather it recommends that decisions for late-phase response be made using a site-specific optimization process which is a decision-making process intended to take into account the many potential attributes or factors that can affect overall public welfare and restoration decisions" (NCRP [8] (Fig. 20.1).

20.4 Decontamination of Buildings, Infrastructure, and Areas

Decontamination, even of radioactivity, is not a mysterious process. The author has performed decontamination of tools, soils, concrete, metal, and even his own skin, generally using the same "decontamination" techniques that are used to remove dust, grease, or spilled food. Loose surface contamination can simply be wiped up and

Fig. 20.1 Photos from a remediation project, showing some of the activities and equipment used. Compared to the aftermath of a radiological or nuclear attack against a major city, the size of the site and complexity of the project is relatively minor. Author's photos

the biggest challenge is to try not to accidentally whisk it into the air by wiping too vigorously. Loosely adhering contamination can be scrubbed off with soap and water, or with commercially available cleaning solutions; if metal ions are involved then a chelating agent can help out. Contamination fixed in the surface of an object can be chiseled out or removed by removing the surface layers, while for contamination that is distributed volumetrically (e.g. a radioactive liquid that has soaked into and penetrated a porous surface) it might be necessary to decontaminate by removing the volume of concrete (or other material) that contains the radioactivity. But even a process of, say, using a jackhammer to remove a chunk of contaminated concrete floor is a relatively straightforward, if unusual and noisy process.

An RDD will spread contamination over as much as several square kilometers of territory and a nuclear detonation will contaminate an order of magnitude more; if detonated in an urban area this contamination will cover every material and surface that is outdoors. Not only that, but contamination can be sucked into buildings through the ventilation system, through open windows and doors, or even through openings where glass surfaces were destroyed by the blast or by flying objects. This means that decontamination must also include carpeting, linoleum, ventilation ducts (dust and all), a variety of types of painted surfaces, wood, textured synthetics, and so on—all of the materials found in all of the types of buildings, homes, and workplaces of which we can conceive.

A great deal of remediation is likely to consist of simply removing and disposing of contaminated objects. It might not make sense, for example, to spend $500 in materials and technician time to decontaminate a $50 hammer. Or, more likely, it might not make sense to spend $100 million to decontaminate a building that can be rebuilt for $10 million. In other cases it might make sense to gut a building, removing and disposing of the ventilation system, carpets, fixtures, and furniture so that the interior can be refurbished and refinished. Alternately, it might also make sense to design a building ventilation system that, when it senses the introduction of radioactive contamination, can shut down ventilation to avoid spreading it throughout the building. Decisions such as these can have a substantial impact on the eventual cost of decontamination and remediation.

If remediation of buildings is challenging, remediation outdoors is even more so. One reason for this is the fact that so many natural surfaces are porous, making decontamination difficult. The natural world also has a huge variety in types of surfaces both living and non-living, with different physical and chemical properties in each of these categories; consider the different properties of clay, loam, various types of rock, gravel, tree bark, leaves, and so many more, each with different levels of porosity and different depths to which contamination can penetrate.

On top of this, we must remember that the actual world is a combination of natural and artificial so decontamination must be able to address, say, Cs-137 dust on the surface of a glass window, nestled into the rough surface of a brick, adsorbed onto clay minerals, taken up into plants, washed into the storm drains, resting in the cracks between adjacent slabs of concrete, and all of the other variations that can exist indoors and out. Radium, with different physical and chemical properties, will behave differently than will cesium; cobalt and iridium are different yet, and other radionuclides have their own unique properties. It is difficult, if not impossible, to develop a decontamination plan for every physical and chemical form of every radionuclide contaminating every type of surface or material.

Luckily, we have a great deal of experience in decontamination and remediation and in general it makes little sense to develop nuclide- and surface-specific decontamination and remediation plans except for buildings or locations of significant historic or cultural value. Soils can be dug up and replaced with clean soil, asphalt and concrete streets and sidewalks can be removed, building exteriors can be sandblasted, and so forth—all making use of technologies and techniques that have been developed and refined over several decades as well as whatever new techniques might be developed for a specific circumstance. And, in fact, in many cases it might well be less expensive and time-consuming to simply use an existing technology than to spend time (and money) developing a less expensive and more efficient methodology, given the time and expense of research and development.

Finally, as noted above, it will be necessary to decide whether or not some contamination might be better to leave in place. Contamination that is mechanically or chemically attached to the surface of a high-rise building, civic monument, bridge tower, or other tall structure tens or hundreds of meters above street level and that produces little or no exposure inside of a structure will have no health effect on the population; it might not make sense to spend large sums of money—and to occupy a decontamination crew for many weeks or months—to remove contamination that is not causing any harm. This is another of the many decisions that will have to be made in the aftermath of a radiological attack (fallout from a nuclear detonation tends to be relatively short-lived and will decay more quickly than long-lived nuclides that might be used in a radiological attack).

20.5 Remediation Contractors and Oversight

Cleaning up radioactive contamination is fairly straightforward—by and large it is much the same as cleaning up spilled chemicals or any other unwanted materials. And while there are some skills unique to working with radioactivity, these skills have little to do with the actual cleanup. In fact, some aspects of radiation simplify the remediation work compared to chemical spills; unlike chemicals, for example, radiation can be detected fairly easily using relatively inexpensive hand-held equipment, giving workers real-time indications regarding the progress of their work as well as their own contamination levels.

The difficulty with radiological remediation is that radiation and radioactivity are heavily regulated (requiring remediation workers and contractors to have a familiarity with the applicable regulations) and it calls for the use of radiation instruments (requiring the workers and contractors to have a familiarity with the equipment). In addition, radiological remediations tend to be subject to both regulatory and public scrutiny, in large part owing to fears of radiation.

As a result, there are a limited number of contractors with experience conducting radiological remediations—enough to perform the typical level of radiological work that exists. But in the event of a radiological or nuclear attack the amount of such remediation that must be done will vastly outstrip the capacity of experienced and qualified remediation firms. This will present regulatory agencies with a conundrum: delaying remediation and reoccupancy of contaminated areas while waiting for qualified contractors to become available, or allowing work to be performed by contractors who lack substantial experience performing radiological work. If remediation carries on long enough it might also be possible for additional contractors to become qualified to perform radiological remediations, perhaps by first working collaboratively (or as subcontractors) to existing experienced radiological remediation firms to gain the experience needed to work independently.

We also need to consider the customers—property owners, business owners, hospitals, universities, and all the others who own or work in contaminated areas. Consider a small business owner whose shop is inside a contaminated building; no matter how eager the owner is to open for business, they cannot do so until the shop is decontaminated, as well as the streets and sidewalks leading to the shop, the parking lot where customers park, the bus and train lines that bring workers and customers, and so forth. This also raises a question of policy: It is reasonable to expect the government to pay for remediation of streets, sidewalks, train stations, and other public property, but should government also be responsible for decontaminating the private parking lot, the commercial building, and the individual shops and offices inside the building if the cost of doing so is beyond the reach of the individual owners? It is likely that individual property and business owners will either be responsible for the remediation of property that they own. This means that thousands or tens of thousands of individual property owners (personal and commercial both) might be the position of trying to find a contractor to remediate their property; from the contractors' viewpoint they can expect to be inundated with calls from a huge number of people who might not even know what work needs to be done.

There will also be a need for regulatory oversight of remediation efforts to ensure the remedial work is performed properly and to confirm that the affected locations meet the agreed-upon cleanup criteria before they are released for reoccupancy and unrestricted use. In particular, it will be necessary to ensure that decontamination goals are met to avoid exposing workers, residents, visitors, and other members of the public are not exposed to doses of radiation that might pose a long-term risk to their health and to confirm that neither contractors nor owners attempt to save money by doing substandard work, performing poor surveys, or falsifying survey results.

One challenge is that there are not many government regulators with experience in radiological remediation, which might slow the progress of the remedial activities. Picture a single office building with 10 business tenants per floor and 25 floors; if each tenant employs an average of 10 workers then the livelihood of 2500 workers depends on restoring access to the building (requiring remediation of the streets, sidewalks, and mass transit), accessible public areas (lobby, elevators, stairwells, hallways, and restrooms), and a place to work (offices, storage rooms, conference rooms, and reception areas). Returning this building to full use will require overseeing the cleanup of as many as 1000 different rooms in addition to 25 floors' worth of hallways and other public areas. The cleanup itself might require several months or even a few years of work, all of which will require regulatory review of remediation plans, progress reports, survey results, and a final closeout inspection. Simply to return to operation a single building and 2500 workers will require the review and approval of a few thousand plans and reports, thousands of survey results, and multiple inspections and confirmatory surveys of 1000 different rooms (plus hallways, lobby, restrooms, and other public areas). And as this graphic makes clear, an area of just 1 km^2 can contain a great many buildings (Fig. 20.2).

This level of effort will need to be multiplied by the number of affected buildings, not only in the Hot Zone, but in the downwind area contaminated by the contamination plume. Few, if any regulatory bodies have sufficient staffing to perform this level of work; it might be necessary to hire temporary workers with the appropriate skills, to borrow workers from other agencies in unaffected cities and states, or to hire contractors who can be trained to perform the work properly.

Looking at the above, it seems likely that contaminated properties will be owned by people who will need to learn about contamination, remediation, and finding a remediation contractor competent to perform the work. At the same time, there is likely to be a shortage of such contractors, which can lead to either delays or the need

Fig. 20.2 Midtown Manhattan showing a few one square kilometer areas. Note the large number of buildings that could require remediation (or demolition) in the event of an event that contaminated an area of this size, as well as giving an idea the impact on transportation and business

to train new contractors in the appropriate regulatory issues and in good radiological work practices. And this entire process will require regulatory oversight, which will require a means of finding or training regulators capable of competently discharging their responsibilities. All of this will be needed in order for residents to return to their homes, workers to their jobs, and to reoccupy the contaminated areas.

For this reason, the New York City government collaborated with Brookhaven National Laboratory to develop their three-volume *Urban Remediation Response* (URR) project, developing a series of checklists designed to help remediation contractors, property owners, and regulatory oversight personnel to understand their respective responsibilities and perform them properly [9].

20.6 Disposition of Remediation Waste

In the aftermath of the September 11, 2001 terrorist attacks more than 1.2 million tons of material required disposal [10]. Debris from the collapsed buildings was first lifted from the pile of rubble at the site and placed into one of two primary staging areas and, from there, moved to barges that took the debris to an existing landfill on Staten Island where over 50 hectares were set aside for these materials. Some of the debris was structural steel that was later recycled, much was contaminated with asbestos, PCBs, and other harmful substances, and much was simply debris from the buildings. More than 1000 workers were present on the World Trade Center site and several hundred more worked at the landfill. Removing and disposing of the waste from these attacks was a major undertaking—one that would pale in comparison to the magnitude and complexity of remediation following a radiological or nuclear attack.

20.6.1 Overall Flow of Waste

Once remediation begins, rubble will be removed from the areas that were destroyed by the bomb or collapsed by the blast wave. Chemical explosives will typically have a radius of destruction of up to a few hundred meters while for nuclear explosives the "rubble zone" will extend to a distance of as much as 1 km in all direction and the radius of the mass fire might be even larger, while the fallout plume can extend for a few tens of kilometers downwind. It might be necessary to begin removing waste materials at a distance from the site of the attack and to work inwards. Eventually, the removal of waste will include decontamination waste, materials with high levels of fixed contamination, materials that do not warrant decontamination, and even buildings that are demolished because of the high cost of remediation and repair.

After removal from the area being remediated, the waste will likely be placed in one of a number of staging areas. As rubble and waste materials are removed they will need to be sorted (if possible) into radiological and non-radiological streams so that they can be treated appropriately—if possible, only radiological wastes should be sent to radioactive waste storage or disposal areas. These staging areas should be relatively convenient to the area(s) being remediated if possible, as well as to transportation routes to the waste disposal site(s).

From the staging area(s) the waste will be placed into containers (if necessary) and onto trucks, trains, barges, or whatever mode of transportation will be used to transport the waste for final disposition. Once at the waste site (a licensed radioactive waste facility or landfill) the waste will be removed from the transportation and disposed of according to the regulations and procedures governing work at that facility.

At each stage it is important to remember that the waste and contamination might also be evidence for the criminal (and other) investigations taking place. Because of this, no materials should be moved or removed from the scene until they have been released by the lead agency (in the US this will likely be the FBI). In addition, those loading, surveying, and sorting the waste should remain alert for anything that might be of interest to the investigators; this can include unexpected radionuclides, hot spots, source fragments, possible pieces of the weapon and/or vehicles used to deliver it, and so forth.

20.6.2 Removal

Waste will first be removed from the affected are to the staging and sorting area(s), although the manner in which this is done will depend in large part on the size of the site, the size of the material being moved, and the distance to the staging area.

If the distance to the staging area(s) is not too far it might be possible to move materials directly from one area to the other using cranes; for more distant staging area(s) it will likely be necessary to use trucks of some sort.

This is likely to be the most hazardous work, requiring the most care and the most substantial safety precautions—including radiation safety. Hazards include [7]:

- Unstable piles of rubble
- Sharp objects
- Ruptured utility lines
- Moving heavy equipment
- Heat stress and dehydration
- Animal bites (insects, rats, legal and illegal pets, etc.)
- High radiation dose rates from sources or pieces of sources
- High contamination levels and resuspended contamination
- Smoke and fumes (materials removed from the World Trade Center site continued to burn or to flare into flame when removed from the pile for several weeks after the towers collapsed)
- Asbestos and other hazardous building materials
- Toxic chemicals
- Lifting and moving heavy loads
- Biological and infectious hazards from bodies buried in the rubble.

Because of this, it will be important for all persons removing material from the site of the attack and affected areas to wear appropriate protective equipment, including radiation dosimeters and to be trained in standard radiation safety practices.

As materials are moved to the trucks, picked up by crane, or otherwise removed from the affected area they should be surveyed for radiation and contamination and marked or sorted appropriately; this will help with further sorting at the staging area(s) and will help reduce work and radiation exposure for those working with the waste later.

20.6.3 Initial Staging and Sorting

The first stop for most waste will be the staging area(s); this is where the waste will initially be examined and surveyed to determine its forensic value (in case this was missed at the scene) as well as its contamination status. The staging area(s) is the first intermediate step between the area that was attacked or affected by the attack and the final disposal; this is where the waste is first examined closely and characterized for whichever contaminants and hazards are felt to pose a concern.

A small amount of material might be suitable for decontamination and re-use, but most will likely be sorted into "contaminated" and "uncontaminated" piles, possibly with further characterization for asbestos, hazardous chemicals, and other threats because many of these must be handled differently in order to comply with appropriate regulations. It is possible that waste disposal regulations might be changed to permit the disposal of multiple waste forms into the same landfill to reduce the amount of sorting and facilitate sorting and loading efficiency as well as to consolidate shipments and destinations.

All personnel working in the staging area will be exposed to a number of possible hazards, similar to those noted in the previous section. However, radiological hazards in the staging area should be better-defined and lower than during initial removal operations because the waste should at least partially characterized during removal.

In the staging area(s), the waste will also be placed into packages (if possible) for transportation; this might be 1 m^3 bags, 2.5 m^3 metal boxes, 200 l drums, or other acceptable containers. They will then be secured for transport and will be taken to the destination. There might also be very large pieces of debris from building collapse; these might require cutting or breaking into smaller sizes for transportation and burial.

20.6.4 Transportation

Packages of waste will be loaded onto appropriate transportation; likely train or barge, although trucks might be used as well if the city is not close to waterways or train lines. Whatever the mode of transport, the proper procedures must be followed and appropriate precautions taken.

Radioactive materials transportation entails too many regulatory and practical details to do justice to the subject here; those who transport radioactive materials on a regular basis are required to take a class on the subject and to receive regular refresher training in order to maintain their knowledge level. Accordingly, this will provide a brief summary rather than a comprehensive discussion of the topic.

Both national (e.g. US DOT 49 CFR 171, 172, and 173 [12]) and international (e.g. IAEA [3]) bodies have developed standards for radioactive materials transportation and shipping. Many nations' regulations are consistent with international standards, making it possible to make some general statements that should be generally applicable.

Radioactive materials transportation requirements include:

- *Acceptable packaging* sufficient to protect and contain the radioactive materials and to keep them from escaping in the event of mishandling or an accident.
- *Proper labeling and placarding* to notify and warn others of the level of risk posed by the package or vehicle contents.
- *Acceptable radiation dose rates* to ensure the package and its contents do not pose a risk to workers or to the members of the public.
- *Securing and safeguarding the package* against loss or theft.

There will likely be contractors responsible for this process and it will be their responsibility to ensure that it is done properly, meeting all relevant regulatory requirements and taking all appropriate precautions, under the supervision of the appropriate regulatory agencies.

20.6.5 Final Disposal

The remediation waste will be shipped for disposal at a licensed radioactive waste disposal facility such as the one shown in Fig. 20.3. This might be an existing facility if the nation that was attacked has an operating facility of this type already in operation; if not then it might be necessary to construct one. In the aftermath of the radiation accident in Goiânia, the Brazilian government constructed a radioactive waste facility to house about 3500 cubic meters of radioactive waste that was generated during remediation activities.

Whether it is an existing facility or one constructed for this purpose, it will have to meet certain criteria to isolate the waste for a period of time; this period will depend on the effective half-life of the radionuclide(s) present; short-lived radionuclides will decay reasonably quickly and will not require a long isolation period while longer-lived nuclides (e.g. Am-241) might require safe isolation for a few to several millennia. At the other extreme, waste from a nuclear attack is likely to decay to safe levels fairly quickly.

Characteristics of an effective new waste facility might include factors such as [15]:

- *Depth to groundwater*—a deeper water table means that contamination will have to travel further and will take longer to reach an aquifer; also minimizes the risk of groundwater infiltration

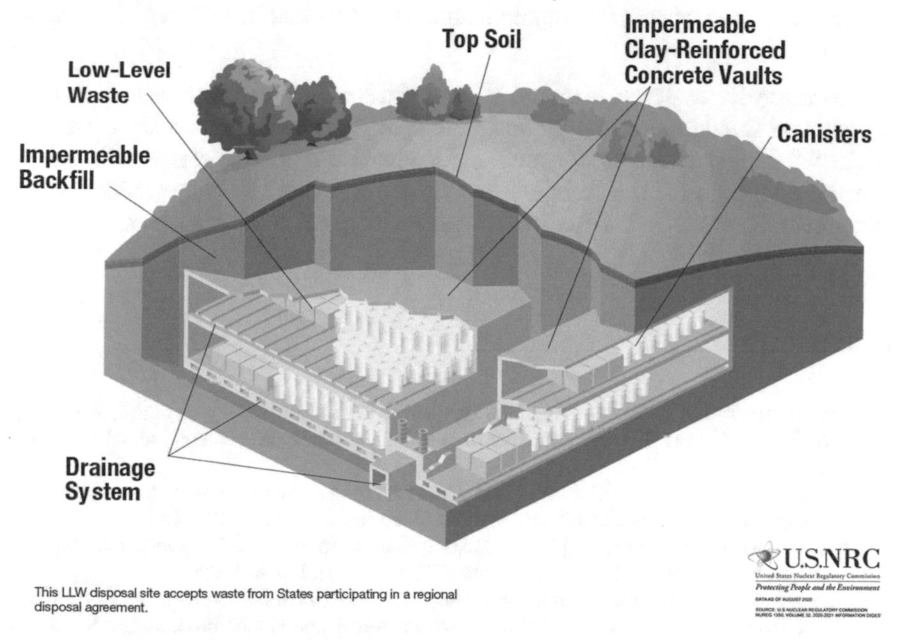

Fig. 20.3 Cross-section of a low-level radioactive waste disposal facility showing the various features designed to prevent leakage of radioactivity into the environment. US Nuclear Regulatory Commission (https://www.nrc.gov/reading-rm/doc-collections/infographics/low-level-radioactive-waste-disposal.png) Accessed December 7, 2020

- *Proximity to population centers*—a greater distance means that any releases to the air or groundwater will take longer to reach the population center and will be more diffuse when they arrive
- *Proximity to surface waters (rivers, lakes, streams,* etc.)—a greater distance to surface waters reduces the impact of leaks on the ecology of local water bodies and those who use them
- *Proximity to transportation routes*—closer distance leads to easier access and lower transportation risk
- *Proximity to cultural, historic, recreational, and other important locations*—to minimize the possible impact on these locations in the event of an accident or release
- *Weather and climate*—ideally located in an arid location to minimize infiltration of water into the facility
- *Local geology (for buried facilities)*—stable, unfractured, rock or soil with low porosity and low permeability delays water infiltration and the migration of contaminants

- *Lack of natural disasters*—volcanic eruptions, major earthquakes, tsunamis, tornadoes, and other catastrophic natural disasters can have a negative impact on facility integrity.

The waste disposal facility should ideally be licensed by the appropriate regulatory agency and should be staffed by qualified, competent radiation workers. Arriving waste will likely be handled the same as any other radioactive waste, with the caveat that the aftermath of a radiological or nuclear attack is likely to produce larger quantities of building and demolition debris than typical waste streams at present.

References

1. Chen S, Tenforde T (2010) Optimization approaches to decision making on long-term cleanup and site restoration following a nuclear or radiological terrorism incident. Homel Secur Aff VI(1)
2. Elcock D, Klemic G, Taboas A (2004) Establishing remediation levels in response to a radiological dispersal event (or "dirty bomb"). Environ Sci Technol 38(9):2505–2512
3. International Atomic Energy Agency (2018) Regulations for the Safe Transport of Radioactive Material, Specific Safety Requirements No. SSR-6 (Rev 1). IAEA, Vienna
4. Karam P (1999) To remediate or not: a case study of Co-60 contamination at the Southerly Waste Water Treatment Plant, Cleveland Ohio, USA. In: Andersson K (ed) Proceedings, VALDOR (Values in Decisions on Risk) symposium, Stockholm Sweden, 13-17 June, pp 173–180
5. Keeney R (1994) Decisions about life-threatening risks. New Engl J Med 331(2):193
6. Leone D (2020) Seattle Cesium spill should be clean in about a year, NNSA nonproliferation chief says. Nuclear Security and Deterrence Monitor, August 28
7. Lorber M, Gibb H, Grant L, Pinto J, Pleil J, Cleverly D (2007) Assessment of inhalation exposures and potential health risks to the general population that resulted from the collapse of the World Trade Center towers. Risk Anal 27(3):1203–1221
8. National Council on Radiation Protection and Measurements (2014) Decision making for late-phase recovery from major nuclear or radiological incidents, report no 175. NCRP, Bethesda MD
9. New York City Department of Health and Mental Hygiene (2013) Urban remediation response: contractors' oversight manual, decision-making toolkit, and property owners' manual for environmental cleanup response to a radiological dispersal device. New York City DOHMH, New York City
10. Nordgren M, Goldstein E, Izeman M (2002) The environmental impacts of the world trade center attacks: a preliminary assessment. Natural Resources Defense Council, New York City
11. US Department of Homeland Security (2008) Planning guidance for protection and recovery following radiological dispersal device (RDD) and improvised nuclear device (IND) incidents. 73 FR 45029-45048. US Government Printing Office, Washington DC
12. US Department of Transportation (2020) Code of federal regulations title 49, Parts 171, 172, and 173. Washington DC. 2020
13. US Environmental Protection Agency (2017) PAG Manual: Protection Action Guides and Planning Guidance for Radiological Incidents (EPA-400/R-17/001. USEPA, Washington DC
14. US Environmental Protection Agency (2019) Federal guidance report no. 15: external exposure to radionuclides in air, water and soil. US EPA, Washington DC
15. US Nuclear Regulatory Commission (1989) Regulating the disposal of low-level radioactive waste: a guide to the nuclear regulatory commission's 10 CFR part 61. Office of Nuclear Material Safety and Safeguards, U.S. Nuclear Regulatory Commission, Washington DC

Chapter 21
Reoccupying the Hot Zone

As remediation is completed and the debris is removed from the affected areas it will be possible to collapse the Hot Zone boundaries inwards and to begin reoccupying the area; this will involve repairing or rebuilding structures, repairing damaged utility lines and other infrastructure, and completing all the other actions needed for the affected area(s) to again become habitable and useful. But until this work is completed, the affected area(s) will remain unoccupied, the business and residents located in that area will remain displaced, and the city will lose not only the area itself, but also the jobs, business, and revenues that are no longer being generated.

In addition to the obvious (clearing rubble, remediation, demolishing buildings that cannot be repaired or decontaminated cost-effectively, and so forth), restoring utilities will be another prerequisite for reoccupying the Hot Zone. As with other activities, this will likely occur in stages, as areas are cleared of rubble and contamination, but it must be done in order to facilitate remediation and reconstruction activities and to make reoccupancy possible.

Even after the area is ready for reoccupancy there might still be difficulties persuading the public to return to the area for short visits or for longer periods of time due to widespread fear of radiation. For many, the lure of home will be sufficient to draw them back; for others, even this might not suffice. Government might need to not only determine and confirm the standards required for reoccupancy to occur, but also to find ways to persuade businesses and residents to return.

This is a subject that has been studied over the years, with lessons learned from events of a variety of magnitudes. The National Council on Radiation Protection and Measurements [7] provides an excellent summary of many of the points to be addressed in this chapter and served as a reference for much of what follows.

P. A. Karam, *Radiological and Nuclear Terrorism*,
Advanced Sciences and Technologies for Security Applications,
https://doi.org/10.1007/978-3-030-69162-2_21

21.1 Preparing for Reoccupancy

Remediation standards are a good first step towards developing a set of requirements for reoccupancy to occur, but they are only a first step. As noted elsewhere, the site will pose multiple risks, all of which will need to be addressed before people can return, and reoccupancy should not be allowed until each of these hazards has been satisfactorily addressed. That being said, there is no reason that reoccupancy needs to wait until every bit of work has been completed; it is not uncommon for reoccupancy to take place over a period of years. In fact it is not uncommon for reoccupancy of evacuated areas to take place over extended periods of time—this has been the experience not only in the areas surrounding Chernobyl and Fukushima, but also in areas hit by massive earthquakes, storms, wildfires, volcanic eruptions, and the like. While, in general, reoccupancy requirements will vary from place to place and from disaster to disaster, there are some general comments that are likely to apply to most, if not all, such efforts.

21.1.1 General Safety

The general safety requirements are among the most obvious. Damaged buildings, for example, must be stabilized and repaired (and pass inspection to confirm their safety) before they can be occupied; fires must be extinguished, utility lines (natural gas, water, electricity, sewer) must be repaired and inspected, and much more. Seeing to general safety includes clearing away the rubble and evicting rats, feral animals, insects, and other pests, as well as anything else that can cause harm.

Depending on the nature of the attack and the extent of the damage it might not be readily apparent when the general safety concerns have been resolved; in the aftermath of the September 11, 2001 terrorist attacks in New York City fires burned underground for several weeks following the attack and those working on the rubble pile told the author that pieces of debris removed from the pile (pieces of structural steel, for example, that penetrated deep into the pile) would be glowing red-hot or would burst into flame when removed from the pile and exposed to the air.

Another concern would be the crater left by a nuclear or chemical explosive, especially if it extends below the water table or is filled by ruptured water or sewer lines. The crater need not be filled in before reoccupancy can begin, but it will at least need to be stabilized so that it presents little or (preferably) no risk to those returning to the area.

The majority of these hazards are likely to be addressed during debris removal and remediation; in many nations the regulations regarding workplace safety—even on construction sites—are as stringent as those for the general public, if not more so. However, it is still prudent to conduct a thorough walkthrough and safety inspection prior to opening each area and each building to ensure the safety of those working, visiting, and living in the area.

General hazards will always exist; the goal should not be to eliminate general hazards altogether but, rather, to attend to the most dangerous (so that nobody need worry about death) and the most common (to minimize those who are injured) so that the public can enter the area(s) safely.

21.1.2 Hazardous Materials (Non-radiological) Safety

In the aftermath of the terrorist attacks on September 11, 2001 a number of governmental agencies began sampling the air and the debris. Contrary to initial expectations, it was found that both of these media were contaminated with a large number of hazardous materials, including asbestos, jet fuel, cleaning chemicals, depleted uranium (used as counterweights for elevators), lead, chromium, nickel, PCBs, dioxins, furans, silica dust, volatile organic compounds, and many others; there were also combustion products from mostof these in the air and that settled out on the ground and on the debris pile [2, 5]. The aftermath of a radiological attack is likely to pose similar non-radiological health risks.

Awareness of and the ability to sample for similar risks was much lower at the time of the nuclear attacks in Hiroshima and Nagasaki; in addition, many of the hazardous materials found in the World Trade Center buildings in 2001 were not present in Hiroshima in 1945, so we have little direct information about hazardous material risks in the aftermath of a nuclear attack. It seems reasonable to assume that the mass fires will incinerate many (if not most) organic compounds, and that the high temperatures of the fireball will vaporize the great majority of harmful materials within a radius of a few to several hundred meters from the scene of the detonation. At greater distances, however, the aftermath of a nuclear attack should be assumed to pose similar non-radiological hazards as those following a radiological attack. However, the wider radius of destruction (including broken glass) means that many more buildings are likely to be affected by a nuclear attack with a concomitant increase in the total volume of potentially hazardous materials to clean up.

Before reoccupancy can take place, it will be necessary to perform extensive sampling to confirm that all of these hazardous materials have been cleared from the remaining buildings. This will involve sampling for asbestos, chemical residues, and other contaminants as well as looking for physical hazards (e.g. broken glass, damaged structures, and the like).

21.1.3 Radiological Safety

Radiation dose rate and contamination levels will depend on many factors, including the amount of radioactivity involved, the amount of time that has gone by and the half-life of the isotope(s) used, the amount of area contaminated, the types and energies of radiation emitted, as well as factors affecting the amount of radioactivity that can be resuspended.

External radiation exposure from gamma-emitting radionuclides is fairly straightforward, as discussed in the first two chapters of this book; but calculating the dose from internal radioactivity and from resuspended radioactivity can be challenging. This is where software models can be helpful; in particular, a series of computer codes called ResRad (which stands for "residual radioactivity") includes one program developed for this purpose—RESRAD-RDD [8].

There will likely be a cycle that develops for the radiological aspects of determining remediation limits:

1. Collect radiological data
2. Develop dose models
3. Determine appropriate cleanup standards
4. Conduct remediation to achieve cleanup standards
5. Collect radiological data and perform new modeling.

A crucial aspect of this cycle will require radiological sampling and surveying; this will include air sampling, contamination monitoring, and measurements of both dose and dose rate to monitor the progress of the cleanup as well as to confirm the accuracy of the computer models and as input to help refine further modeling.

As areas are checked by regulatory agencies and released for reoccupancy it will be possible for workers and business owners to move back into their places of business and for tenants to return to their homes as soon as streets, sidewalks, and other access routes are cleared and reopened.

It is also worth noting that it might not make sense to reoccupy every affected area; some areas might not be worth the cost of remediation and it might make more sense to cordon them off and to relocate the people and businesses from those areas rather than to remediate or demolish and rebuild the affected properties. In a city, for example, with low occupancy rates for apartment and office buildings, it is likely to cost less to refurbish underutilized buildings and the surrounding infrastructure than to tear up concrete and asphalt, demolish heavily contaminated buildings, and dispose of the debris as radioactive waste.

21.1.4 Organizing the Decision-Making

The 2014 NCRP report also includes a discussion on the process of developing recommendations and making decisions on whether or not they should be implemented and how that should be done. As shown in the graphic, NCRP suggests a Decision-Making Team comprised of representatives from all levels of government from the local to the national, taking input from stakeholders as well as from a Recovery Management Team; the Recovery Management Team, in turn, receives input and suggestions from a number of Technical Working groups, each assigned a different problem or problem area.

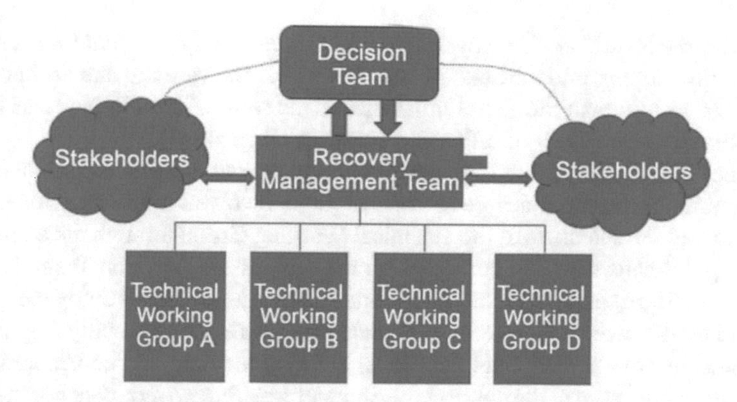

Fig. 21.1 Example of an organizational approach for the recovery process. With permission of the National Council on Radiation Protection and Measurements, [7] https://NCRPonline.org

In this structure, a Technical Working Group will develop a scientific or technical approach to addressing a problem or decision that will be presented to the Recovery Management Team; the Recovery Management Team will share this with various stakeholders and will incorporate their feedback into the final recommendation(s) presented to the Decision Team. The Decision Team will begin with the best scientific and technical advice, stakeholder concerns, and will add in economic, social, political, and other considerations to arrive at a decision that, with luck, will be reasonable and that will be acceptable to the greatest number of people (Fig. 21.1).

21.2 Persuading Residents and Businesses to Return

In their 2014 report referenced above, the NCRP states that "For a severely affected community, the objective of the long-term recovery is to return the community to an acceptable new normality in the most expedient manner possible, with a goal to re-establish and sustain the local economic and social viability." Remediation is only a part of this process, but even the most complete remediation will not help to "re-establish and sustain the local economic and social viability" if nobody is willing to return to the area.

The public's fears of radiation and the stigmatization of people and areas affected by radiological and nuclear accidents in the past suggests that, even after an area meets reoccupancy standards it might remain underpopulated, or even vacant, for some time. The effects of this will be discussed in greater detail in the following chapter; for now, let it suffice to note that the government might need to offer more than data and reassurances and that it may require some sort of incentives to persuade many to return home. In New York City, for example, most of Lower Manhattan was evacuated in the aftermath of the attacks against the World Trade Center, with residents reluctant to return to a location that had seen so much tragedy. Among other incentives, the New York City government subsidized occupants' rent for six months for anybody moving into the area; in less than a year the vacancy rate dropped from about 30% to less than 5% [3]. Similar incentives were offered to business tenants to encourage reoccupancy of office and commercial space [4].

As the NCRP notes, this will require an acceptance of these decisions on the part of the public, requiring a degree of trust in those involved in the decision-making process, from the scientists of the Technical Working Group through the administrators and politicians who will comprise the majority of the Decision Team. Unfortunately, scientific experts, administrators, and politicians no longer enjoy the public's trust and confidence to the extent they once did, making it especially important to communicate not only the final decisions, but also the data and considerations on which they were based to the public. Unfortunately, neither government nor the scientific and technical community have a good track record in communicating with the public convincingly in the aftermath of radiological and nuclear accidents, making it important to identify spokespersons that the public is likely to trust [1, 6].

21.3 Lessons Learned from Historic Incidents

The 2014 NCRP report included summaries of several relevant accidents and other events and included a summary of lessons learned from these experiences. These lessons are summarized here.

1. Although radiological and nuclear incidents are not common, enough have occurred to have accumulated valuable experience and to have an understanding of the issues involved. NCRP feels it is not unlikely to assume that many of the long-term recovery issues from an RDD or IND attack will be similar to those following these past incidents, albeit with site-specific issues.
2. While resilient cities recover from disasters more rapidly than those that lack resilience, it is difficult to build resilience after an attack or accident has taken place; community resilience is best built before it is needed. Resilience depends on a number of economic, political, and cultural factors, as well as the desire of the community itself to recover.
3. Openly communicating the particulars of an event is crucial to recovery and can help to address issues of social stigmatization that have been seen in the aftermath of several radiological and nuclear accidents. Secrecy can lead to a loss of trust in the authorities that can exacerbate this stigmatization.
4. Involving all stakeholders in the decision-making process tends to lead to greater acceptance of those decisions, whether these decisions involve the locations of waste staging areas, remediation criteria, waste shipment, and other factors involved in remediation and reoccupancy.
5. It is likely that any large-scale event will require the development of new technologies, practices, and/or equipment to address site-specific problems

and conditions, in addition to taking advantage of technologies and practices developed in response to past events that might be applicable.
6. Waste management appears to be a nearly universal problem in past radiological and nuclear events. Thus, it seems appropriate to include waste management in pre-event recovery planning.
7. It will be necessary to plan for long-term environmental and public health monitoring, including infrastructure, food production and processing, food chains, and more. In addition to collecting valuable information, such monitoring can also give the public confidence that these issues are being addressed properly.

References

1. Becker S (2004) Emergency communication and information issues in terrorist events involving radioactive materials. Biosecur Bioterror Biodef Strategy Pract Sci 2(3):195–207
2. Biello D (2011) What was in the World Trade Center Plume? Scientific American, Sept 7 2011. Available online at https://www.scientificamerican.com/article/what-was-in-the-world-trade-center-plume/. Accessed 25 Nov 2020
3. Fong W (2002) Low rents, cash grants lure tenants back to ground zero. Wall Street J
4. Fuerst F (2006) The aftermath of the 9/11 attack in the New York City office market: a review of key figures and developments. Munich Personal RePEc Archive
5. Lorber M, Gibb H, Grant L, Pinto J, Pleil J, Cleverly D (2007) Assessment of inhalation exposures and potential health risks to the general population that resulted from the collapse of the World Trade Center towers. Risk Anal 27(3):1203–1221
6. National Council on Radiation Protection and Measurements (2001) Management of terrorist events involving radioactive material, report no 138. NCRP, Bethesda MD
7. National Council on Radiation Protection and Measurements (2014) Decision making for late-phase recovery from major nuclear or radiological incidents (report no 175). NCRP, Bethesda MD
8. Yu C, Cheng J, Kamboj S, Domotor S, Wallo A (2009) Preliminary report on operational guidelines developed for use in emergency preparedness and response to a radiological dispersal device incident, DOE/HW-0001, ANL/EVS/TM/09-1. US Department of Energy, Oak Ridge TN

Chapter 22
Long-Term Public Health, Economic, and Societal Concerns

Remediation, recovery, rebuilding, and reoccupancy in the aftermath of a radiological or nuclear attack might take from years to decades to accomplish. But the effects of such an attack might last for years or decades longer still. The atomic bombings in Japan, for example, occurred more than 75 years before this writing, but survivors and their descendants continue to be studied to glean more information about the long-term health effects of their radiation exposure. The nation of Japan, in spite of three-quarters of a century of recovery and healing, still suffers remnants of the psychological trauma inflicted by those attacks, exacerbated by the later accidents in Tokaimura and Fukushima. Similarly, the consequences of Chernobyl continue to be felt in Ukraine and Belarus after more then three decades. Even a much smaller event such as the radiological accident in Goiania Brazil continued to resonate for years after the event itself had been resolved.

Past experience suggests that the city—and likely the nation—attacked with radiological or nuclear weapons will continue to suffer the aftereffects for a considerable length of time after the cleanup ends. It behooves governments to develop plans for addressing this more protracted phase following an attack.

22.1 Public Health Concerns

Radiation's deterministic effects are largely resolved within weeks or months (although it is not uncommon for health effects to continue to manifest themselves for a few to several years following exposure [5, 8, 13]). By and large, however, the acute radiation injuries and other health effects should be either healed within a year or so. Other health effects, however, can continue for years or decades and radiogenic cancer usually does not begin to appear for at least five years post-exposure and can continue to appear for several decades afterwards. The timeline for acute radiation injury is well-known as is the probability and approximate latency period

P. A. Karam, *Radiological and Nuclear Terrorism*,
Advanced Sciences and Technologies for Security Applications,
https://doi.org/10.1007/978-3-030-69162-2_22

for radiation-induced cancer; these are health effects for which public health officials can plan in advance. The health effects of non-radiological agents (e.g. inhaled dust, vapors, and fumes) are less understood but, as the experience with World Trade Center survivors has shown, they can also continue having an impact for at least a few decades.

22.1.1 Disease Registry

Even before the nuclear weapons were used against Japan the scientists and physicians of the Manhattan Project realized that their use would expose large numbers of people to doses of radiation that could vary from inconsequential to lethal; they also realized that medical science lacked an understanding of what to expect in both the long-term and the short-term aftermath of a nuclear detonation. The Japanese physicians (as well as Allied physicians who arrived after the Japanese surrender) saw and treated tens of thousands of patients in the days and weeks following the explosions and, in November 1946, President Truman ordered the formation of the Atomic Bomb Casualty Commission (ABCC) to learn about the medical impact of nuclear detonations.

The ABCC was an American organization that worked with Japanese physicians and government; in 1945 the ABCC was superseded by a joint US-Japanese venture, the Radiation Effects Research Foundation (RERF). The RERF has organized its work into six primary research programs and additional research areas:

- Life Span Study
- Adult Health Study
- Immunology
- Cytogenetics
- Atomic-bomb dosimetry
- Children of Atomic-bomb Survivors Study
- In utero study
- Radiation Biology
- Molecular Epidemiology
- Statistics.

Two aspects of RERF's work should be of interest to public health officials in cities at high risk of being subject to a radiological or nuclear attack: the manner in which it is structured and the work it has been doing, and what can be learned from the work already performed as it applies to the possible aftereffects of a future attack.

The structure of the RERF is not overly complex and is described on the Foundation's website (https://www.rerf.or.jp/en/). While it does not make sense to staff up a foundation of this sort anticipating that it might be needed at some point in the future, it might be reasonable to at least establish the procedures and policies so that, if needed, this "skeleton" can be fleshed out with staff. In addition, it should be possible to establish a framework that will also lend itself to other events involving

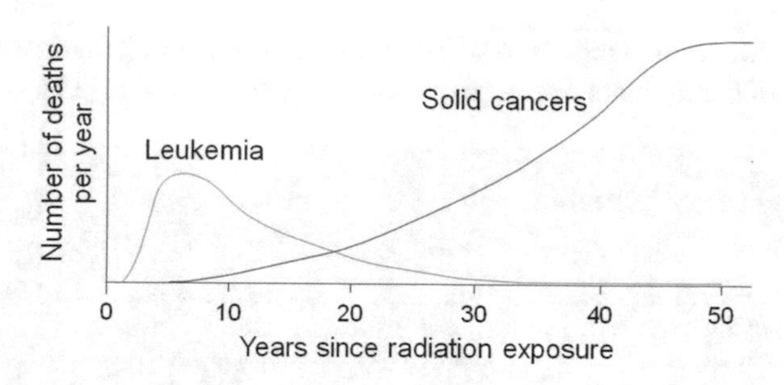

Fig. 22.1 Excess cancer deaths per year following the nuclear attacks against Japan. Note that leukemia deaths peak about 5–10 years post-exposure while solid cancers continue increasing for about 45 years and then level off. We might expect to see a similar effect in the aftermath of a nuclear terrorist attack. Used with the permission of the Radiation Effects Research Foundation

environmental contaminants (e.g. asbestos from a steam pipe explosion or building collapse) to develop and maintain proficiency in using the tracking system and to make it easier to activate when needed.

More importantly, public health officials should consider studying the RERF's findings, possibly arranging for periodic briefings, to remain current in this area to help anticipate not only long-term public health needs, but also to anticipate the size and nature of various programs (e.g. treatment and compensation programs to those who become ill as a result of radiation exposure) that might need to be set up in the aftermath of an attack (Fig. 22.1).

22.1.2 Radiological Injury and Dose Reconstruction

The cancers that tend to appear first are cancers of the blood such as leukemia [5], appearing about 5–10 years after exposure, followed by solid tumors beginning about a decade or so after exposure and continuing to rise each year thereafter. However, in a city of, say, five million people we would expect over one million to die of cancer in this same time frame (in the US). Thus, even an additional 50,000 cancer fatalities over a 50-year period would not make for a substantial increase in the annual cancer rates within the city.

Having said that, if the government or insurance companies compensate those whose cancer is determined to be due to radiation exposure from the attack then it will be necessary to evaluate each claim to determine the likelihood that each cancer was (or was not) due to the attack. This will involve work at all levels of government and might need to be contracted out if there are insufficient governmental resources (especially trained dose assessors) to perform the work. An American radiation exposure compensation program, for example, required the full-time efforts of more than 200 trained radiation safety professionals to perform dose assessments on nearly 53,000 claimants over a period of nearly two decades [11].

How many cancers might result from a nuclear attack?

As calculated in Chap. 9, the radiation dose from an IND will be approximately 9 Gy at a distance of 1 km from the device. Using the inverse square law (and neglecting the shielding effects of large masonry buildings and terrain), the dose at larger distances will drop with the inverse square of distance. To simplify calculations, we can assume a population density of about 10,000 people per square kilometer. Using these values and assuming a lifetime fatal cancer incidence of 5% for each Sv of exposure we can fill in the Table 22.1 (please note that these numbers are approximate and are more precise than can be justified, were these calculations to be presented and used formally, as they are based on simplistic assumptions). Interestingly, because the area (and population) increases with the square of the radius while the dose drops with the square of the radius, the population dose and estimated future cases of cancer remains roughly constant at each radius.

22.1.3 *Non-radiological Health Effects*

We must also remember the likelihood of exposure to non-radiological contaminants; it might be necessary to include non-radiological factors (e.g. concrete dust, asbestos) and the diseases they cause to the disease registry. For example, the September 11, 2001 disease registry includes a long list of conditions felt to be caused by exposure to one or more of the contaminants present in Lower Manhattan in the weeks and months following the attack; these include a large number of cancers in addition to

Table 22.1 Calculations of radiation-induced cancers following an IND attack

Radius (km)	Area (km^2)	Population	Dose (Sv)	Population dose (person-Sv)	Future cancer fatalities
1	3.1	31,000	9	279,000	0[a]
2	12.6	126,000	2.25	283,500	14,175
3	28.3	283,000	1	283,000	14,150
4	50.3	503,000	0.56	281,680	14,084
5	78.5	785,000	0.36	282,600	14,130
Total					**56,539**

[a]Every person within a radius of 1 km is likely to die in the initial explosion or from radiation exposure. Thus, there are few or no expected cancer fatalities from this population

conditions such as asthma, interstitial lung diseases, upper airway hyperreactivity, chronic obstructive pulmonary disease, and more [7, 17].

It is tempting to simply combine the lists of cancers that can be induced by radiation with those that can be caused by non-radiological airborne contaminants such as silica dust or asbestos. Unfortunately, there is a great deal of overlap between these lists; in the event of a radiological or nuclear attack it might be necessary to evaluate the probability of causation for both radiogenic and non-radiogenic cancers to determine whether or not a particular cancer, or any other condition, is compensable. This will add complexity to the evaluation process and might delay compensation decisions. It might also require additional staff as there are few professionals who are competent to assess both radiological and non-radiological exposure and probability of injury from either or both of these.

22.1.4 Mental Health Issues

Long-term public health concerns must also include mental health concerns. There have been a number of recognized post-September 11 mental health issues (e.g. [3, 12]), as well as observations made by the World Health Organization (WHO) among those forcibly relocated following the Chernobyl accident, including:

- Acute stress disorder
- Adjustment disorder
- Anxiety disorder (not otherwise specified)
- Depression (not otherwise specified)
- Dysthymic disorder
- Generalized anxiety disorder
- Major depressive disorder
- Panic disorder
- Post-traumatic stress disorder (PTSD)
- Substance abuse
- Suicide.

Many of these were also noted by the World Health Organization among survivors of the Chernobyl reactor accident [18] and several are consistent with observations made by the author and his teammates in the aftermath of the Fukushima accident [1, 2]. Given that the post-Chernobyl mental health concerns continued for at least 20 years after that accident, public health officials should assume that they will remain a concern for at least this long following a radiological or nuclear terrorist attack, as well as lasting trauma of the terrorist attack itself (PTSD, hyperarousal, relationship difficulties, etc.; [15]). It is also worth noting that the WHO also found evidence of mental health problems among children of those who were evacuated, including children who were not born until after the accident.

Thus, the health effects of a radiological or nuclear attack can last beyond the generation involved in the attack; this, in turn, will require trained personnel to attend to those affected as well as the financial resources required to pay for not only the treatment, but also for whatever compensation the government deems appropriate.

22.2 Economic Concerns

The financial impact of a radiological or nuclear attack is likely to be massive. The financial impact in New York City of the September 11, 2001 attacks has been estimated as being as high as several hundred billion dollars [14], with a national impact that was even higher. This is of the same order of magnitude as Loeb and Zimmerman's estimates of the economic impact of a radiological attack [20]. and when the nation's response to these attacks—wars in Afghanistan and Iraq, the creation of the Department of Homeland Security, an increased military presence throughout the world, increased funding for law enforcement, security, and other response capabilities, and the expense of all of these—the cost to the United States alone rises to several trillion dollars, with additional costs borne by all of the nations involved in these wars on either side.

Consider some of the costs directly associated with a radiological or nuclear attack:

- Value of the buildings destroyed
- Value of the furnishings, sales goods, and other materials in the buildings
- Cost of damaged and destroyed infrastructure
- Cost of destroyed and damaged public and private vehicles
- Value of lives lost (estimated at $7 million per life [16])
- Loss of rental income by landlords
- Loss of business income
- Loss of sales and business tax revenue
- Reduction in tourism and tourism-related sales as well as the jobs, incomes, and the resulting tax revenue
- Loss of income tax revenue from killed or displaced residents
- Salaries (including overtime) of emergency and medical responders
- Renovation and reconstruction of damaged and destroyed buildings
- Remediation of contaminated areas
- Radioactive waste transportation and disposal
- Construction of waste staging area(s)
- Medical care for injured
- Long-term compensation (and possibly medical care) for compensable injuries and illnesses.

These costs have very little to do with the rest of the state or province the city is in or with the nation itself. For example, both regional and national resources will almost certainly be sent to assist with the response and both regional and national governments will also suffer from lost tax revenues. Beyond that, however, the costs are more intangible and more difficult to quantify.

- The cost of the wars in Afghanistan and Iraq were mentioned earlier; there is no way to know in advance if another large-scale terrorist attack would lead to another war, to multiple wars, or no war at all.
- The September 11, 2001 attacks led to large-scale changes in the American government; we cannot predict whether or not a future attack would lead to a similar set of changes.
- Any attack involving weapons of mass destruction would likely cause large-scale societal disruption, but the magnitude and economic cost of this disruption cannot be predicted.
- The September 11 attacks resulted in the cessation of all air travel within the United States and between the US and the rest of the world for more than two days and caused a roughly 30% reduction in air travel within the US during the month of September [9]; we cannot predict the impact of a different attack on a different date using different weapons and tactics.
- The September 11 attacks led to the creation of new capabilities within all levels of government in the United States, including military, law enforcement, medical, and other response agencies. However, this capability already exists and would not need to be created again (although existing capabilities might be augmented).

These lists are unlikely to be exhaustive, but they also become more difficult to predict or to quantify. As one example; the value of a lost human life was determined in 2005 for American citizens [16], but is that value appropriate for citizens of other nations nearly two decades later? And, of course, this assumes that we can predict the numbers of lives that would be lost and the nations in which these losses would occur in the event of one or more hypothetical wars in undetermined locations at an uncertain time in the future.

That being said, in seems reasonable to assume that, in addition to causing more damage and a higher economic cost, a nuclear attack would likely cause a more dramatic impact in the nation attacked and overseas. For example, if nuclear forensic techniques are able to identify the nation in which the fissionable material was produced it is not unreasonable to assume that this nation would be held responsible in part or in whole for the attack and its aftermath. This is likely to lead to a retaliation that could include bombing, invasion, regional war, or a more general war. A radiological attack, on the other hand, being far less deadly and causing much less damage, might lead to actions more analogous to the aftermath of the attacks on September 11, 2001 (e.g. a more localized war or series of attacks by the nation that was attacked and its allies against those found to be responsible for the terrorist attack).

22.2.1 Economic Impact on the Neighborhood(s) Attacked

Consider a person who has a daily routine; on their way to work they board a bus at the neighborhood bus stop after picking up a cup of coffee at the local delicatessen.

When they get off the bus near their office they get a breakfast sandwich from the food cart halfway between the bus stop and their office building, eating it as they walk. For their coffee break perhaps they return to the food cart, and there are a half-dozen places they alternate between for lunch. Their routine includes their local grocery store, another half-dozen or so places where they eat dinner on the weekends, a pub, and so forth. Most of us have some sort of a routine that we follow—a limited set of places where we shop, eat, drink, and socialize.

Now consider this same person in the aftermath of a radiological attack that leaves their office, food cart, and workplace inside the Hot Zone and in which the plume spreads as far as their home neighborhood. The local businesses near the office and near home will be forced to close for a time, our worker will have to move to a new place, as will the business at which they work. They will develop a new routine that includes different stores, different eateries, different … everything. Unless they are insured or supported by loans or grants many of these places will go out of business; even with some form of support, though, their customers cannot return until the area has been remediated and reoccupied. But by that time, their former customers will all have fallen into new routines that involve other businesses. Even after the Hot Zone and contaminated areas are released for reoccupancy, some—perhaps most—of the former residents will have settled into their new places and might be reluctant to return, resulting in a loss of rent on the parts of the landlords and other property owners. Even if the businesses (including office and apartment buildings) are given financial support, fears of radiation and the new habits their customers have picked up are likely to delay the economic recovery of the Hot Zone and all the other areas affected by the attack. Not only that, but each business employs workers (as well as owners); each business that closes means a loss of livelihood for these people, changing them from taxpayers to public assistance recipients.

In the case of a radiological attack, the affected is likely to be relatively limited in extent, and can still involve up to several square kilometers in area. In the event of a nuclear attack the affected area is likely to be on the order of several tens of square kilometers with a concomitantly larger population of affected people and businesses. As the world has seen with the global COVID-19 pandemic, local businesses can suffer greatly when they are cut off from their customers and it is not unlikely that the same would happen in the Hot Zone and other affected areas in the aftermath of a radiological or nuclear attack.

It is also important to realize that remediation and reoccupancy takes time, and during that time it is entirely possible that the businesses that do not shut their doors will simply relocate elsewhere with their former places of business remaining vacant until someone else can be persuaded to sign a lease. But in the time that passes—possibly several years or even a few decades—many landlords might have their banks foreclose on mortgages, themselves going out of business. Without going into excessive detail, it is easy to see how the economic impact of even a relatively minor attack can snowball, with the extent of that impact rippling outwards into the larger community over time. And this is not a purely hypothetical example; this is what was seen in Lower Manhattan in the aftermath of the 2001 terrorist attacks as well as in other cities subjected to terrorist attacks [14].

If an attack involves only relatively short-lived radionuclides, such as those used in medicine, it might be possible to wait for them to decay to stability; as a rough rule of thumb, in ten half-lives the amount of radioactivity drops by a factor of 1024. Iodine-131, for example, has a half-life of about 8 days; it will take about three months to go through ten half-lives of decay. Thus, an area attacked using I-131 could be considered "off-limits" for three months or more. The economic impact of such an attack is likely to be much lower than for an attack using long-lived radionuclides (ten half-lives of Cs-137, for example, is three centuries) because remediation will not be necessary and because the local businesses will likely be able to weather being shut down for a few months; at the same time, workers will still be unemployed, and both workers and business owners are likely to require assistance of some sort to remain solvent for the period of time until the radionuclide decays to acceptable levels, with the corresponding economic cost. Even an attack using a relatively short-lived radionuclide can lead to a relatively large area being placed "off-limits" for weeks or months and can cost society tens or hundreds of millions of dollars in lost wages, lost business, lost tax revenues, and long-term loss of tourism and related business.

22.3 Psychological and Societal Concerns

If the experiences of Chernobyl, Three Mile Island, Goiania, and other radiological and nuclear accidents are considered then it seems reasonable to assume that a radiological or nuclear attack will have long-lasting impacts on individuals and on society as a whole.

22.3.1 Psychological Trauma

Twenty years after the accident the World Health Organization conducted a wide-ranging study of people living in areas affected by the Chernobyl accident, including those who were evacuated from contaminated areas [19]. In a review of WHO's findings, the IAEA summarized the report's most significant points, including several relevant to this section [18].

- *Poverty, "lifestyle" diseases now rampant in the former Soviet Union and mental health problems pose a far greater threat to local communities than does radiation exposure.*
- *Relocation proved a "deeply traumatic experience" for some 350,000 people moved out of the affected areas. Although 116 000 were moved from the most heavily impacted area immediately after the accident, later relocations did little to reduce radiation exposure.*
- *Persistent myths and misperceptions about the threat of radiation have resulted in "paralyzing fatalism" among residents of affected areas.*

- *Ambitious rehabilitation and social benefit programs started by the former Soviet Union, and continued by Belarus, Russia and Ukraine, need reformulation due to changes in radiation conditions, poor targeting and funding shortages.*

Alongside radiation-induced deaths and diseases, the report labels the mental health impact of Chernobyl as "the largest public health problem created by the accident" and partially attributes this damaging psychological impact to a lack of accurate information. These problems manifest as negative self-assessments of health, belief in a shortened life expectancy, lack of initiative, and dependency on assistance from the state.

"Two decades after the Chernobyl accident, residents in the affected areas still lack the information they need to lead the healthy and productive lives that are possible," explains Louisa Vinton, Chernobyl focal point at the UNDP. "We are advising our partner governments that they must reach people with accurate information, not only about how to live safely in regions of low-level contamination, but also about leading healthy lifestyles and creating new livelihoods."

The WHO report was conducted two decades after the Chernobyl accident and it was clear that hundreds of thousands of people were still suffering from the psychological trauma of the event and that this was likely to continue for some time to come. The psychological trauma of Chernobyl has already lasted for a generation.

This should not be surprising; the Hibakusha (Japanese atomic bomb survivors) were stigmatized for several decades following the nuclear attacks against Japan with resulting in long-standing mental health concerns. In her 2013 thesis, University of Oregon MS student Michelle Heath [6] noted that 24% of Hibakusha responding to a survey felt the need to hide their status in order to avoid discrimination, that the group reported diminished marital prospects, with similar effects noted following the 2011 Fukushima accident.

Social stigmatization and lasting psychological trauma have been observed in the aftermath of the majority of large-scale radiological and nuclear events; it seems reasonable to assume that the aftermath of an attack would be no different. In fact, things might be even worse in this regard, considering that there would also be the added trauma of the terrorist attack. In one publication, Neria and others [12] note a significant level of post-traumatic stress disorder (PTSD) in the aftermath of the terrorist attacks on September 11, 2001 along with accompanying depression, anxiety, and substance abuse—similar to what WHO noted in their 2006 report about Chernobyl survivors, and to what Heath described regarding the Hibakusha in her thesis. Neria et al. report on telephone surveys shortly after the attacks that revealed 44% of adult Americans surveyed reported symptoms of PTSD, with the incidence dropping to 5.8% six months later. It must be stressed that these surveys were of all Americans, not only those in the cities that were attacked; 5.8% of all Americans means that more than more than 16 million Americans remained traumatized six months after these attacks.

At somewhat later times—up to five years after the attacks—various studies showed a few to several percent of the general population continued to exhibit symptoms of PTSD, and nearly 20% of emergency responders [4]. Interestingly, while rates of PTSD dropped with the amount of time that elapsed following the attacks, those who were directly involved—emergency responders, persons in the buildings that were struck, those who lived or worked in Lower Manhattan, etc.—reported higher rates of PTSD over longer periods of time, possibly owing to on-going stresses in their lives.

All of this suggests that the trauma from terrorist attacks and the trauma from radiological and nuclear disasters is long-lasting, and it seems reasonable to speculate that the trauma from a radiological or nuclear terrorist attack might be even more traumatic for a longer period of time. This, in turn, suggests that public health agencies and mental health professionals should consider the need to prepare to provide mental health services to residents—primarily in the city attacked, but nationally as well—for at least a generation following the attack. These services should include care for PTSD and its accompanying depression, anxiety, substance abuse, and related ills [10].

22.3.2 Societal Trauma

We know that individuals can be traumatized, causing them to act in ways that are not typical for them. Since societies are comprised of individuals, it is reasonable to speculate that any event that traumatizes a large number of individuals might cause their entire society to exhibit behavior that is contrary to the society's "normal" and to exhibit the societal equivalent of trauma.

Consider once again the aftermath of the terrorist attacks against the United States on September 11, 2001. Americans were individually traumatized by an attack against their national capital and their largest and most iconic city. Individuals in New York City and Washington DC—the cities that were attacked—were most traumatized; they could see and smell the smoke, New Yorkers could see the gap in the Manhattan skyline, many of them had family and friends who were killed or injured. For months—years in some cases—afterwards residents of these cities were hyper-aware of their surroundings, of people who appeared to be engaged in suspicious activities, and many were careful to maintain emergency supplies (e.g. candles, "go" bags, food, etc.) on hand in the event of another attack.

But much of this behavior extended far beyond residents of New York City and Washington DC. Hospitals out to several hundred kilometers from these cities cancelled elective surgeries to be able to accommodate possible victims from the September 11 or future attacks. The author worked at a hospital located 500 km from New York City and knew some physicians who joined the military Reserves as a result of the attacks, and others even further away traveled to Washington and New York or sent food, money, and supplies to these cities to help victims and emergency responders alike.

At the same time, Americans were nervously watching even small aircraft taking off or landing at regional airports, they were looking out for anybody who appeared to be Muslim or from the Middle East (and reporting such persons to the police), they were even refusing to share a bus, train, or airplane with anybody wearing a turban, a beard, or having a darker complexion.

But the most visible impact was seen on a national level. The attacks help to bring the nation together, and a nation filled with frightened people demanded that their politicians take actions to ensure that those who had perpetuated the attack were punished so that another such attack would not take place again. At the time of the attacks, the American military had not fought in a major conflict for more than a decade; since the attacks, American troops have been engaged in combat in both major and minor conflicts for nearly two decades with (as of this writing) no end in sight. The collective trauma of the September 11 attacks also led directly to a major reorganization of the American government, the expenditure of hundreds of billions of dollars in national security programs, disagreements with many allies of the US, a reappraisal of many American international priorities, substantial changes in American security policies and practices, and much more. Some of these changes (e.g. increased airport security precautions and the invasion of Afghanistan) were likely inevitable; others (e.g. the invasion of Iraq, waging the Global War on Terror, the law known as the Patriot Act) may have been a result of the trauma suffered by the American people.

The aftermath of a terrorist attack using radiological or nuclear weapons in any nation is likely to induce a national trauma comparable to that which afflicted the United States following the terrorist attacks of September 11, 2001. And, as happened in the US, such an attack could very well cause a nation to come together in the short term, calling for or authorizing actions felt to enhance security. In the long term, however, it is entirely plausible to feel that, as also happened in the US, the stress and tension of the national trauma might exacerbate divisions within the nation as various groups process the information and emotions surrounding the attack in their own manner. It is also reasonable to expect that all of these effects might last for decades. Thus, the psychological and societal effects of such an attack might well last as long, or even longer, as the radiological effects.

References

1. Becker S (2011) Learning from the 2011 Great East Japan Disaster: insights from a special radiological emergency assistance mission. Biosecur Bioterror 9:394–404
2. Becker S (2013) The Fukushima Dai-Ichi accident: additional lessons from a radiological emergency assistance mission. Health Phys 105(5):455–461
3. Centers for Disease Control and Prevention (2002) Psychological and emotional effects of the September 11 attacks on the world trade center—Connecticut, New Jersey, and New York, 2001. Morb Mortal Wkl Rep 51(35):784–786
4. Chiu S, Niles J, Webber M, Zeig-Owens R, Gustave J, Lee R, Prezant D (2011) Evaluating risk factors and possible mediation effects in posttraumatic depression and posttraumatic stress disorder comorbidity. Public Health Rep 126:201–209
5. Hall E, Giaccia A (2019) Radiobiology for the radiologist, 8th edn. Lippincott Williams and Wilkins, New York
6. Heath M (2013) Radiation stigma, mental health and marriage discrimination: the social side-effects of the Fukushima Daiichi nuclear disaster. MA Thesis, Department of International Studies, University of Oregon

7. Herbert R, Moline J, Skloot G et al (2006) The World Trade Center disaster and the health of workers: five year assessment of a unique medical screening program. Environ Health Perspect 114(12):1853–1858
8. International Atomic Energy Agency (2002) The radiological accident in Gilan. IAEA, Vienna
9. Ito H, Lee D (2005) Assessing the impact of the September 11 terrorist attacks on U.S. airline demand. J Econ Bus 57:75–95
10. Joyner E, van der Hoorn K, Platt A, Rubin M, Shvil E, Neria Y (2016) Placing collective trauma within its social context: the case of 9/11 attacks. In: Ataria Y, Gurevitz D, Pedaya H, Neria Y (eds) Interdisciplinary handbook of Trauma and culture. Springer, Amsterdam, pp 325–337
11. National Institute of Industrial and Occupational Safety and Health. NIOSH radiation dose reconstruction program. https://www.cdc.gov/niosh/ocas/default.html. Accessed 25 Nov 2020
12. Neria Y, DiGrande L, Adams B (2011) Posttraumatic stress disorder following the September 11, 2001 terrorist attacks: a review of the literature among highly exposed populations. Am Psychol 66(6):429–466
13. Radiation Effects Research Foundation. Late effects on survivors: from several years after the atomic bombings to the present day. https://www.rerf.or.jp/en/programs/roadmap_e/health_effects-en/late-en/. Accessed on 25 Nov 2020
14. Rose A and Blomberg S (2010). Total economic consequences of terrorist attacks: insights from 9/11. Peace Econ Peace Sci Public Policy 16(1). https://doi.org/10.2202/1554-8597.1189. Accessed 25 Nov 2020
15. Verger P, Dab W, Lamping D, Loze J, Deschaseaux-Voinet C, Abenhaim L, Rouillon F (2004) The psychological impact of terrorism: an epidemiological study of posttraumatic stress disorder and associated factors in victims of the 1995–1996 bombings in France. Am J Psychiatry 161:1384–1389
16. Viscusi W (2005) The value of life. Discussion Paper No. 517. Harvard Law School, Cambridge MA
17. Wisnivesky J, Teitelbaum S, Todd A et al (2011) Persistence of multiple illnesses in World Trade Center rescue and recovery workers: a cohort study. Lancet 378(9794):888–897
18. World Health Organization (2005) Chernobyl: the true scale of the accident. https://www.who.int/news/item/05-09-2005-chernobyl-the-true-scale-of-the-accident#:~:text=According%20to%20the%20Forum's%20report,danger%20to%20their%20health%20from. Accessed 14 Nov 2020
19. World Health Organization (2006) Health effects of the chernobyl accident and special health care programmes. In: Bennett B, Repacholi M, Carr Z (eds) Report of the UN chernobyl forum expert group "Health." World Health Organization, Geneva
20. Zimmerman P, Loeb C (2004) Dirty bombs: the threat revisited. Def Horiz 38:1–11

Concluding Thoughts

At the very least, this book should make it clear that this is a hugely complex topic. Some aspects—the physics of nuclear weapons, the health effects of radiation exposure, medical treatment for radiation sickness, and other scientific and technical areas—are supported by a robust body of knowledge and experience. Other aspects—the best way to manage the scene of a nuclear attack, the most effective use of personnel and equipment for nuclear interdiction, the most efficient operation of a Community Reception Center, and so forth—remain largely hypothetical since they have never been practiced in real life and may never be. In spite of this (or perhaps because of this), there is a huge body of literature on all aspects of preventing and responding to these events.

This book, like all books, represents a snapshot in time; it is an imperfect summary of what is known at the time it was written. With so much literature it is impossible to read and incorporate it all; with so little experience of actual events it is impossible to do more than guess in many areas; even when these guesses are well-informed by research and experience, they remain our best guesses as of 2021.

P. A. Karam, *Radiological and Nuclear Terrorism*,
Advanced Sciences and Technologies for Security Applications,
https://doi.org/10.1007/978-3-030-69162-2

Glossary

Absorbed dose The amount of energy deposited in a material by ionizing radiation. The units of absorbed dose are the gray (Gy) and the rad.

Activity The rate at which radiation is emitted by radioactive material. The units of activity are the becquerel (Bq) and the curie (Ci).

Airborne radioactive material Any radioactive material dispersed in the form of dusts, fumes, particulates, mists, vapors, or gases.

Airborne radioactivity area A room, enclosure, or area in which airborne radioactive materials exist in concentrations high enough that an individual present in the area without respiratory protective equipment could inhale enough radioactivity to reach a dose of 30 mrem (300 mSv) in one week of work.

Annual limit of intake (ALI) The limit for the amount of radioactive material that can be ingested or inhaled in one year without exceeding a radiation dose limit.

As low as reasonably achievable (ALARA) Making every reasonable effort to maintain exposures to radiation as far below the regulatory dose limits as is practical, consistent with the purpose for which the licensed activity is undertaken, taking into account the state of technology, the economics of improvements in relation to the state of technology, the economics of improvements in relation to benefits to the public health and safety, and other societal and socioeconomic considerations, and in relation to utilization of licensed or registered sources of radiation in the public interest.

Background radiation Radiation from cosmic sources; naturally occurring radioactive materials, including radon, except as a decay product of source or special nuclear material, and including global fallout as it exists in the environment from the testing of nuclear explosive devices.

Becquerel (Bq) Equal to one disintegration or transformation per second.

Calibration The determination of

(1) the response or reading of an instrument relative to a series of known radiation values over the range of the instrument, or

(2) the strength of a source of radiation relative to a standard

P. A. Karam, *Radiological and Nuclear Terrorism*,
Advanced Sciences and Technologies for Security Applications,
https://doi.org/10.1007/978-3-030-69162-2

Committed dose equivalent The dose equivalent to organs or tissues that will be received from an intake of radioactive material by an individual during the 50-year period following the intake.

Contamination The deposition of radioactive material in any place where it is not desired (units = dpm).

Controlled area Any area where access is controlled for the purpose of protecting individuals from exposure to radiation and radioactive material. "Controlled area" is synonymous with "restricted area."

Curie (Ci) A unit of activity. One curie is that quantity of radioactive material that decays at the rate of 3.7×10^{10} decays per second.

Decay chain (Series) A series of isotopes resulting from the decay of a parent nuclide and its subsequent radioactive daughters to an ultimate stable form.

Derived air concentration (DAC) The concentration of a given radionuclide in air, which, if breathed without respiratory protection for a working year of 2,000 h under conditions of light work, will give a radiation dose of 5 rem (50 mSv) to the whole body.

Derived air concentration-hour (DAC-hour) The product of the concentration of radioactive material in air, expressed as a fraction or multiple of the derived air concentration for each radionuclide, and the time of exposure to that radionuclide in hours. A person may receive up to 2000 DAC-hours of exposure annually.

Dose A generic term that refers to a person's exposure to radiation.

Dose limits The permissible upper bounds of radiation doses established in accordance with regulations.

Dose meter An instrument designed to measure the dose rate of ionizing radiation—usually displayed in mR/hr or mGy/hr.

Dose rate Absorbed dose delivered per unit time (mR/hr or mGy/hr).

Dosimeter An instrument used to detect and measure accumulated radiation exposure.

Element A category of atoms having the same number of protons and the same chemical properties.

External dose That portion of the dose equivalent received from any source of radiation outside the body.

Extremity Hand, elbow, arm below the elbow, foot, knee, and leg below the knee.

Gray (Gy) The SI unit of absorbed dose. One gray is equal to 100 rad.

Half-life (radiological) The amount of time that is required for a radioactive substance to lose ½ of its activity.

Half-value layer (HVL) The amount of material required to reduce the dose rate from a radiation source by a factor of 2.

High-radiation area Any area, accessible to individuals, in which radiation levels could result in an individual receiving in excess of 1 mSv (0.1 rem) in 1 h at 30 cm from any source of radiation or from any surface that the radiation penetrates.

Incident Any unexpected event involving radiation or radioactivity with the potential for the spread of radioactive contamination, skin contamination, uptake of radioactive materials, exposure to elevated radiation levels, loss of radioactive material, and so forth.

Internal dose That portion of the dose equivalent received from radioactive material taken into the body.

Isotope An atom of the same atomic number (containing the same number of protons) but with a different number of neutrons in the nucleus (different atomic mass)—can be stable or radioactive.

Licensed material Radioactive material received, possessed, used, transferred, or disposed of under a general license or specific license issued by a regulatory agency.

Monitoring The measurement of radiation, radioactive material concentrations, surface area activities, or quantities of radioactive material, and the use of the results of these measurements to evaluate potential exposures and doses.

NORM A Naturally Occurring Radioactive Material—material that is naturally radioactive due to containing naturally occurring radioactive isotopes.

Nuclide An atom characterized by the number of protons in its nucleus AND its energy level [ex: Tc-99 m is a different nuclide than Tc-99, exhibiting a different half-life and different decay energies due to its existing in a different nuclear excitation (metastable) state].

Occupational dose The dose received by an individual in the course of employment in which the individual's assigned duties involve exposure to sources of radiation, whether in the possession of the licensee, registrant, or other person. Occupational dose does not include doses received from natural background radiation, as a patient from medical practices, from voluntary participation in medical research programs, or as a member of the public.

Quality factor (Q) The conversion factor used to derive dose equivalent from absorbed dose.

Rad The unit of absorbed dose. One millirad equals 0.001 rad.

Radiation Alpha particles, beta particles, gamma rays, x-rays, neutrons, high-speed electrons, high-speed protons, and other particles capable of producing ions. For purposes of this manual, ionizing radiation is an equivalent term.

Radiation area Any area, accessible to individuals, in which radiation levels could result in an individual receiving a dose equivalent in excess of 0.05 mSv (0.005 rem) in 1 h at 30 cm from the source of radiation or from any surface that the radiation penetrates.

Radiation source Any radioactive material or any radiation equipment.

Radioactive material Any solid, liquid, or gas that emits radiation spontaneously.

Radiation worker Any person who, because of their work, has the potential to receive a dose of 100 mrem (1 mSv) or greater in one year from working with or around radiation or radioactive materials. All radiation workers must receive training prior to commencing work with radiation or radioactive materials.

Radioactivity The property of certain nuclides of spontaneously emitting particles or gamma radiation following orbital electron capture, electron emission, isometric transition, nuclear rearrangement, or spontaneous fission.

Radionuclide A radioactive nuclide.

Rem The unit of any of the quantities expressed as dose equivalent. The dose equivalent in rem is equal to the absorbed dose in rad multiplied by the quality factor (1 rem = 0.01 Sv).

Respiratory protective equipment An apparatus, such as a respirator, used to reduce an individual's intake of airborne radioactive material.

Restricted area Any area, access to which is limited by the licensee for the purpose of protecting individuals against undue risks from exposure to sources of radiation.

Sealed source Radioactive material that is permanently bonded or fixed in a capsule or matrix designed to prevent release and dispersal of the radioactive material under the most severe conditions that are likely to be encountered in normal use and handling.

Sievert The SI unit of any of the quantities expressed as dose equivalent. The dose equivalent in sievert is equal to the absorbed dose in gray multiplied by the quality factor (1 Sv = 100 rem).

Survey An evaluation of the radiological conditions and potential hazards incident to the production, use, transfer, release, disposal, or presence of sources of radiation. When appropriate, such evaluation includes, but it is not limited to, tests, physical examinations, and measurements of levels of radiation or concentrations of radioactive material present.

Very high radiation area An area, accessible to individuals, in which radiation levels could result in an individual receiving an absorbed dose in excess of 5 Gy (500 rad) in 1 h at 1 m from a source of radiation or from any surface that the radiation penetrates.

Whole body For purposes of external exposure, head, trunk (including male gonads), arms above the elbow, or legs above the knee.

Printed in the United States
by Baker & Taylor Publisher Services